S0-BZG-843

HAWKS
AT A DISTANCE

HAWKS
AT A DISTANCE
IDENTIFICATION OF MIGRANT RAPTORS

JERRY LIGUORI

Foreword by
PETE DUNNE

PRINCETON UNIVERSITY PRESS   PRINCETON AND OXFORD

Copyright © 2011 by Princeton University Press

Published by Princeton University Press, 41 William Street,
Princeton, New Jersey 08540
In the United Kingdom: Princeton University Press, 6 Oxford Street,
Woodstock, Oxfordshire OX20 1TW
nathist.princeton.edu

All Rights Reserved

Library of Congress Cataloging-in-Publication Data

Liguori, Jerry, 1966–
    Hawks at a distance : identification of migrant raptors / Jerry Liguori ; foreword by Pete Dunne.
        p. cm.
    Includes bibliographical references and index.
    ISBN 978-0-691-13558-8 (hardcover : alk. paper) — ISBN 978-0-691-13559-5 (pbk. : alk. paper)
1. Falconiformes—North America—Identification.   2. Falconiformes—Flight—North
America—Pictorial works.   3. Falconiformes—Migration—North America.   I. Title.
    QL696.F3L538 2011
    598.9'44097—dc22   2010032990

British Library Cataloging-in-Publication Data is available

This book has been composed in Sabon LT Std text with Gill Sans Std display.
Printed on acid-free paper. ∞
Printed in China
10 9 8 7 6 5 4 3 2 1

In memory of Jon Jon Stravers,

whose passion for raptors was unsurpassed

and who loved and cherished the natural world

Contents

Foreword

It's all about "the distance." It's always been about "the distance." Ever since that first hominid ancestor fixed front-seated eyes on some dinosaur descendent, the challenge has been vaulting distance and getting across an invisible line. The line that defines the border between "Too Far" and "Close Enough."

Too Far. A region marked by frustration where birds lie beyond the limits of ambition and skill.

Close Enough. A happy place where birds, and our designs upon them, meet and become one.

Well, Close Enough just got closer (in fact, you are holding the keys to the kingdom in your hands) and Too Far just got pushed clear to the horizon. Jerry Liguori's new book has redefined the art of hawk watching and pushed field guides to new heights.

And distances!

It's been a long journey, the road to Far. In fact, it's taken more than a hundred years for field guides to reach a point where our ability to detect birds and our ability to identify them lie within a wingspan of each other. But the human compulsion to pin names to birds goes back even farther—back to the time when the primary tool of bird study was the shotgun.

Before the twentieth century the line between Close Enough and Too Far was much, much closer than it is today. Descriptions of birds applied to birds held in the hand (which is about as close as Far can get) and from the standpoint of identifying birds in the field they were about as far from useful as an instructional tool can be.

Take, for example, this species description lifted from Frank M. Chapman's "Handbook" entitled *Birds of Eastern North America*, first published in 1895.

"Upper parts fuscous, more or less marginated with ochraceous or rufous; region below the eye black; ear coverts buffy; wings as in the ad; upper surface of the tail barred with grayish, under surface barred with ochraceous-buff; underparts cream-buff or ochraceous-buff, streaked, spotted or barred with black."

Anybody guess juvenile Peregrine? Thought not. But a new age in bird identification was just on the horizon. In 1906 Chester A. Reed published a handy, shirt-pocket-sized guide to the birds whose utilitarian genius lay in depicting as well as describing his subjects. The Chester Reed Guides featured painted illustrations of birds in natural settings. It was immensely popular and it even worked!

Kind of. To a point. A point limited not only by distance but a perceptual misconception that led to a fundamental design flaw. The somewhat muted illustrations found in Reed's guides attempted to portray birds as they might appear in the field to the unaided eye. The problem was Reed didn't get it. Didn't appreciate the fact that as birds get farther away, images don't get fainter or fuzzier. In fact, just the opposite happens. Contrasts sharpen. Patterns not visible before tighten and become manifest.

A string of dots running down the breast fuse into streaks. Translucent spots near the tips of outer flight feathers form a pale, crescent that bisects the tip of the wing.

What Reed did was take a bird in the hand and make it look like it was being viewed through a dirty window.

It wasn't until 1934 that a field guide got the principles right. In the middle of the Great Depression a 25-year-old "student of birds" published a book that depicted birds as a viewer might see them using the finest optics of the day.

The artist was Roger Tory Peterson. He was the proud owner of a pair of 4x Le-Maire binoculars—the state of the art in optical technology back when gas was ten cents a gallon and Shirley Temple made her film debut.

If you're interested in seeing what birds looked like using first-generation binoculars, open the pages of a first edition Peterson field guide and study the plates. You will discover that most of the illustrations are black and white. In the entire book there are only three color plates (and all the Eastern warblers are crammed onto one of these). Birds are depicted in rigid profile, viewed from the side. Tiny arrows direct viewers to each bird's distinguishing traits, or "trademarks of nature," as Peterson called them.

Better known today as "field marks."

This was the book those proto-hawk watchers took with them onto ridge tops and out to the tips of peninsulas in the 1930s and 1940s. But once again and just like Reed's, the illustrations fell a step behind reality and need. The "Peterson Principle" relied heavily upon plumage-based field marks and its focus was birds in the bush (not in the hand)—birding's great leap forward in 1934. The illustrations in his guide simply didn't correspond to the distant, flying forms that hawk watchers were finding in the sky—birding's high frontier in 1934.

No aspersion upon Peterson, whose contribution to bird identification was seminal and profound. It's just that birding's frontier is never static. As soon one barrier is vaulted, the challenge and frontier moves on.

But one thing had not changed. This was the guiding principle that was the foundation of Peterson's leap. The need to depict birds *as they really appear*. Here Peterson was smack on target and subsequent writers and illustrators have done little more than move field guides closer to that ideal just as they moved the limits of bird identification closer to the horizon.

Take Richard Pough and Don Eckelberry's *Audubon Guides*, first published in 1946. The illustrations, all of them in color, depicted birds in assorted, lifelike poses. Some turned this way; some that. Rigid was out; body language and posture were in. Forty years of watching live birds through optics, not over the barrels of shotguns, had brought us to this new level of awareness and discernment.

While hawks were not shown in flight in the *Audubon Guide*, the illustrations of perched birds were excellent—a credit to Eckelberry's discerning eye and artistic skills.

OK, I lied. What I should have said was that the color "plates" depicted perched raptors. The written species accounts *were* accompanied by a black-and-white, line-art illustration of hawks in flight. But like Peterson, artist Eckelberry got lost in the details—showing plumage characteristics that were apparent at close range but not at a distance.

Eckelberry's sense of shape and proportion also fell short. But only twenty years after Peterson's first guide (not to mention the founding of Hawk Mountain Sanctuary—the raptor "classroom in the clouds") the improvement in how flying raptors were perceived and depicted was marked and manifest.

Clearly, Eckelberry had observed a flying hawk or two. So too, artist Arthur Singer, the illustrator of Chandler Robbins's *Birds of North America,* better known as the "Golden Guide" and published in 1966. In this popular guide, which is still available today, Singer illustrated all the raptors in color and in flight. Then he did something that stepped out of the static realm of mere field identification and into the more dynamic hawk-watching arena.

Capping the individual species accounts, Singer depicted all the raptors, in flight, *on one comparative plate.* For the first time in a popular guide, "students of birds" could compare relative size, plumage, and shape—just as hawk watchers do in the field.

Yes, the birds still showed too much detail. Yes, the shapes still left something to be desired. But it was another step forward and the horizon took another step backward.

In the 1960s and 1970s a tidy swarm of brochures, booklets, and books appeared that focused specifically upon the identification of birds of prey. The crash of raptor populations caused by the widespread and indiscriminate use of DDT had stoked a worldwide interest in raptors. More eyes searching the sky pushed the frontiers of raptor identification farther and farther, and it was during this period that my own fascination with birds of prey began. First in the woodlands and fields behind my parents' home in suburban North Jersey. Later, starting in 1974, at Hawk Mountain, Pennsylvania, and well-known hawk-watching junctions in New Jersey.

Unknown to me, two other young "students of birds" were evolving into ardent raptor fanciers too. One was a Cape May native named Clay Sutton. The other was the son of Yale ornithologist Fred Sibley, growing up in Connecticut. His name was David.

Drawn, ultimately and perhaps inexorably, to the hawk-watching hotbed that is Cape May, New Jersey, the three of us came to craft a book called *Hawks in Flight*—a book dedicated to the identification of migrating raptors. Published in 1987, following on the heels of Bill Clark and Brian Wheeler's *Hawks,* this guide, too, is still in print. Some of the most talented younger field birders raising glasses today have said that *Hawks in Flight* changed the way they looked at birds. One of these, it delights me to say, was Jerry Liguori—the raptor authority whose first book, *Hawks From Every Angle,* was ground breaking and whose new book, *Hawks at a Distance,* is game changing.

Far, now, has no place to run and no place to hide.

So what distinguishes this new photo-driven guide? Precisely that! The wealth of well-chosen images, harvested from the sky, depicting distant birds in flight. Just as optics were the game changer that allowed students of birds to redirect their attention from the bird in the hand to the bird in the bush, then the sky, digital photography has made it possible to snatch those images from the sky and portray them, vividly, accurately, and in a multitude of postures in a book. You can spend thousands of hours studying distant raptors in the effort to anchor a multitude of search images in your mind. Or you can open the pages of *Hawks at a Distance* and "drink this in," as my hawk-watching mentor, Floyd P. Wolfarth, would have encouraged.

But the wealth of images alone does not distinguish *Hawks at a Distance.* Layout, design, and selection are key and here hawk watcher Liguori's experience and genius shines as brightly as Peterson's did back in 1934. Just as Peterson was the master of his subject, Liguori is master of his. In fact, in 2010, you simply cannot find a better hawk watcher than Liguori or a better field guide to birds of prey in their natural element than the one you are holding in your hand, now.

Far. It's not as distant as it used to be. In fact the distance between Close and Far has just been reduced to the width of this book. For birding's horizon to get much farther, *somebody* is going to have to devise a means of compensating for the curvature of the earth. Now that *Hawks at a Distance* is available to today's "students of birds," the world beyond the horizon is about the only place that a raptor that aspires to remain anonymous can hope to hide.

Pete Dunne
New Jersey Audubon's Cape May
Bird Observatory
August 2010

Preface

Beginner birders are fascinated by how experienced birders can identify distant birds with confidence and accuracy. Their first question is usually "what kind of binoculars are those?" thinking the binoculars must be a special model that allows birders to see farther and clearer than any other. It is easy to be impressed by the skill of an accomplished birder, but with enough practice anyone can identify distant hawks. Of course, the farther away a bird is, the harder it may be to identify, and some birds are simply too far away to name. When I began watching hawks, I was drawn to the "specks" in the sky that most birders ignored, and even took pictures of distant raptors with the intent of using them for instruction one day. I wanted to write a book that showed how to identify raptors as they are typically seen in the field. My first book, *Hawks From Every Angle*, accomplishes this and includes close-up plumage information and aspects of identification that are often confused or misinterpreted in the field. To include every aspect of raptor identification in one book would overwhelm readers and birders in the field. Therefore, *Hawks at a Distance* focuses on identification of *distant* raptors in flight. It is like no other identification guide, and fills a niche in the field of identification.

ACKNOWLEDGMENTS

I would like to thank those who helped me to complete this book, especially my wife Sherry, who reviewed the entire book and supplied photographs throughout it. I thank Keith Bildstein and Tony Leukering for their invaluable comments on the text. Thanks also to Aaron "Skippy" Barna, Vic Berardi, Richard Crossley, Kara Donohue, Pete Dunne, Pete Gustas, Julian Hough, Adam Hutchins, Robert Kirk, Michael O'Brien, John Rokita, Deneb Sandack, David Sibley, Brian Sullivan, Clay and Pat Sutton, Dave Tetlow, and Brian Wheeler.

Introduction

Identifying raptors is similar to solving a mystery by piecing together clues. By far, the greatest challenge of raptor identification is naming distant birds. This is because plumage details (and sometimes flight style and structure) can be difficult to judge from afar. Even for the most experienced observers, identifying raptors based on plumage alone can be flawed. With this in mind, imagine how difficult it is to distinguish two species that are practically identical in plumage, like Sharp-shinned and Cooper's Hawks. Even a distinctive plumage trait like the red tail of an adult Red-tailed Hawk can be difficult to view in the field. With practice, **it is more accurate to tell similarly plumaged birds apart by shape and flight-style differences than by coloration.** I identify most of the distant raptors I see based on shape and flight style, using plumage traits only if clearly visible. Remember, there are exceptions to every rule, so almost no field mark is 100 percent exclusive to one species.

At any hawk-watching site, a close-up bird will draw "oohs" and "aahs" from observers. But because most hawks seen in the field are distant, *Hawks at a Distance* is a necessary guide. It focuses on distant birds and discusses traits that are truly useful in the field, leaving out certain others that can only be judged at close range. Be aware that some field marks may be difficult to see in the photos in this book. This is done purposely to show how these field marks truly appear in the field. *Hawks at a Distance* is the first guide that presents birds of unknown identity, pointing out instances when telling age, sex, color morph, or species is impossible, and showing the effect that lighting and molt can have on a bird's appearance. Only migrants that are common

across the United States and Canada are covered. **Most of the photographs in this book were taken during autumn migration from August 15 to November 30. Birds photographed during spring migration from March to late May or otherwise are noted as such.**

I am frequently asked which hawk identification problem I believe is the most difficult. Several come to mind, like telling brown Prairie Merlins from female Kestrels at a distance, ageing Golden Eagles to a specific year, telling a Peregrine Falcon from a Prairie Falcon by silhouette, or telling adult female from juvenile Harriers in spring. However, some aspects of identification are simply impossible due to overlap or similarities in plumages, like telling the race of some Red-tailed Hawks, Merlins, or Peregrine Falcons. Remember, it is enjoyable to simply watch raptors without fussing over their age, sex, or race and that learning hawk identification is an ongoing process.

HOW TO USE THIS BOOK

Hawks at a Distance is designed to show raptors in "real world" settings and help with identification of birds seen with only a quick glimpse. Therefore, the birds throughout these pages are meant to appear distant, unlike most guides that present close-up portraits of birds. *Hawks at a Distance* presents each species in all light conditions. Each color plate is a composite that includes up to six images and a single caption. They are presented this way so that each color plate is viewed in its entirety as the description is read. Throughout the color plates, birds are shown in poses (i.e., soaring, gliding) in which migrants are normally observed. This is intended to stress the shape features referred to in the captions of certain color plates.

The black-and-white plates portray shape characters for each species pointed out by the accompanying captions. These plates are designed so that the reader's eyes wander around the page and take in the entire page quickly. The text is short so that it is easy to remember and so that the book is uncluttered. Key identification traits are in boldface throughout the text, and memorizing these traits will prove to be useful in the field. For example, the traits in boldface for Golden Eagle are "large, dark bird . . . extremely long, somewhat broad wings . . . dihedral when soaring, slightly bowed when gliding . . . slow moving and steady."

One of the keys to identifying raptors is to learn the basic characteristics of each species *well* instead of learning the minutiae for each. For example, recognizing the shapes of each raptor, and understanding how they vary under different circumstances, is much more valuable than knowing the eye color of each raptor. Be aware that certain written or spoken descriptions (particularly regarding wing beat and flight style), and even the same field observation between various people, can be interpreted differently. For instance, I once heard the words "stiff," "choppy," "snappy," and "aggressive" used independently by four separate observers to describe the wing beats of a single bird as it flew by.

TERMINOLOGY

Age Terminology

Some age-related terms can be misunderstood or misinterpreted. The terms "juvenile" and "adult" are fairly straightforward

and easy to understand (see Glossary), but the terms "immature" and "sub-adult" are sometimes unclear. Even some ornithologists disagree on the exact definitions of these terms. Regardless, the terminology throughout this guide is used in a consistent, practical manner.

The term "immature" describes a bird whose plumage is other than adult. Therefore, an immature bird is either a juvenile or a sub-adult. Some argue the term "immature" relates to a bird's sexual maturity regardless of its plumage. For instance, a breeding accipiter in juvenile plumage (which is sometimes the case with females) could be regarded as an adult because it is producing young. However, age terminology herein defines a plumage as it pertains to field identification, not a yearly age or sexual maturity.

The term "sub-adult" describes a bird whose plumage is a distinct plumage altogether from that of an adult or juvenile. Birds in their first adult plumage (that may have none or several juvenile feathers retained) that are essentially identical or appear identical in the field to successive adult plumages are simply "adults." Most raptors reach adulthood after their first molt cycle starting at about one year old. A few species take about two years, while eagles typically take about five years to reach full adult plumage.

Color Morph Terminology

Throughout this guide, the term "dark" is used to describe buteos that are mostly or completely dark on the underside. This includes birds otherwise known as "intermediate morph" and "dark morph." Intermediate and dark-plumaged birds often appear identical to each other at a distance, which is why both are simply classified as "dark" birds throughout this guide. About 90 percent of dark Swainson's, Red-tailed, and Ferruginous Hawks are truly intermediate birds, but are mistaken for dark birds regularly. On the other hand, intermediate plumages of Broad-winged and Rough-legged Hawks are rare, with dark birds being the norm. I refrain from using the terms "morph" and "phase" throughout the text since they can be misleading, and instead use the terms "light" and "dark" to describe a bird's plumage.

The term "intergrade" describes a bird with a plumage that falls between two subspecies (like Taiga and Prairie Merlins) or between a subspecies and a recognized form of a species (like Krider's Red-tailed Hawk, a pale form of the Eastern Red-tailed Hawk). A hybrid is the offspring of two separate species; hybrids rarely occur naturally, but are commonly bred for falconry.

GLOSSARY

Adult plumage - A bird in its definitive plumage.

Axillaries - Wing pits (see Anatomy).

Bib - A dark patch of feathers on the breast.

Buffy - A pale tan coloration.

Carpal - Underwing area at the "wrist" where all the primaries meet (see Anatomy).

Cere - Flesh between the bill and forehead.

Crown - Top of the head.

Dihedral - Wings held above the plane of the body in a "V" position (see Flight Positions).

Dilute plumage - An overall light tan plumage on a normally darker bird.

Flight feathers - Primaries, secondaries, and tail feathers.

Glide - To fly forward with wings pulled in (see Flight Positions).

"Hand" - Consists of all the primaries, the outermost part of the wing (see Anatomy).

Head-on - Eye-level, front profile (see Flight Positions).

Hover - To remain stationary in flight while flapping (see Flight Positions).

Immature - All ages other than adult.

Intergrade - A bird showing traits of two different races or forms.

Juvenile - A bird in its first plumage.

Kettle - A group of birds soaring together.

Leggings - Feathers that cover the legs and sometimes the feet (see Anatomy).

Leucism - Presence of some or all white feathers on a normally darker-plumaged bird. Sometimes referred to as albinism.

Melanism - Presence of dark feathers on a normally lighter-plumaged bird.

Modified dihedral - Position of wings raised at the shoulder and level at the "wrists" (see Flight Positions).

Molt - Replacement of old feathers with new feathers; usually occurs from April through September in raptors.

Morph - Color form.

Nape - Back of neck (see Anatomy).

Patagium - Area between the "wrists" and body along the leading edge of the wings (see Anatomy).

Primaries - Ten outer remiges or "hand" of the wing; the notched outer primaries make up the "fingers" of a hawk (see Anatomy).

Rectrices - Tail feathers.

Remiges - Secondaries and primaries.

Rufous - An orange-rust color.

"Rump" - Feathers covering the bases of the uppertail coverts (see Anatomy).

Scapulars - Feathers along the sides of the back (see Anatomy).

Secondaries - Flight feathers from the

"wrist" to the body making up the base of the wing (see Anatomy).

Sexual dimorphism - Distinct difference between male and female plumages of the same species.

Soaring - Rising in a circular motion with wings outstretched (see Flight Positions).

Stoop - To dive with wings folded (see Flight Positions).

Sub-adult - A bird in plumage (and age) between juvenile and adult.

Subterminal band - The next to last band on the tip of the tail.

Superciliary line - Line of pale feathers over the eye (see Anatomy).

Tawny - Dark yellowish-brown color.

Terminal band - A band at the tip of the tail or wings.

Undertail coverts - Feathers covering the underside of the base of the tail (see Anatomy).

Underwing coverts - Feathers covering the underwing (see Anatomy).

Uppertail coverts - Feathers covering the topside of the base of the tail (see Anatomy).

Upperwing coverts - Feathers that cover the upperwing (see Anatomy).

Wing base - Inner half of the wing from the "wrist" to the body.

Wing-on - An eye-level, side profile (see Flight Positions).

Wing panel - A pale or partially translucent "window" in the primaries (see Anatomy).

"Wrist" - Joint on the leading edge of the wing where the secondaries and primaries meet (see Anatomy).

HAWK MIGRATION

Each spring and autumn, migration occurs across almost all of North America. Raptors move in a broad front, but are known to concentrate along "paths" created by coastlines or mountain ranges. Official hawk migration counts are conducted at many shoreline sites in North America, especially along the Great Lakes, the Atlantic and Pacific oceans, and the Gulf of Mexico. There are popular ridge-top count sites along the Rocky Mountains in the West and the Appalachian Mountains in the East. More than a thousand hawk migration sites are known in North America, and websites exist for many of them. The Hawk Migration Association of North America (HMANA) is an excellent resource for hawk migration sites and migration in general.

HELPFUL HINTS

I cannot imagine how many times I have over-analyzed or second-guessed the identity of a bird only to realize my first impression of the bird was correct. The saying "if it walks like a duck . . . it is a duck" is applicable to all forms of birding, so becoming familiar with the shape and flight style of each raptor species is the key to identifying them. For example, most people can recognize a family member or friend from a long distance because they are familiar with the person's figure or the way the person walks. Learning hawk identification through repeated observations is the same concept.

Sometimes birders hear or read about an identification trait for the first time and then apply it incorrectly in the field. For example, I have seen Cooper's Hawks mistaken for Broad-winged Hawks because they "showed tail bands." Another example is relating a plumage trait that only occurs on an adult bird to a juvenile, or vice versa, or noting the manner in which a bird holds its wings in a certain position when it applies to a different posture. **Some traits apply to only one**

ANATOMY

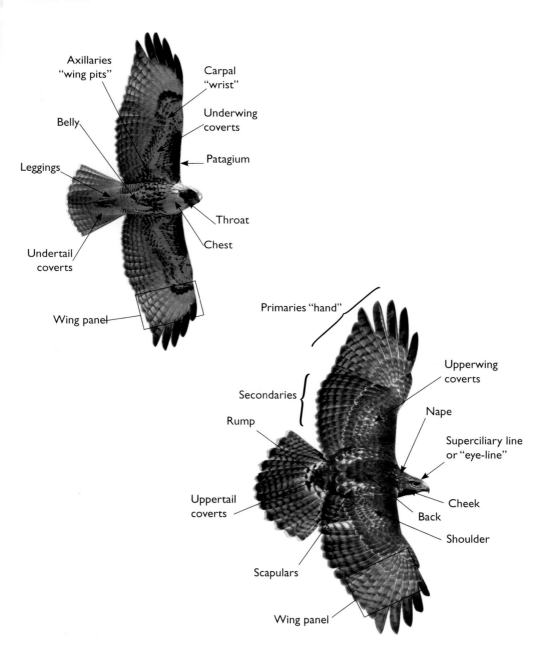

Axillaries "wing pits"

Carpal "wrist"

Belly

Underwing coverts

Leggings

Patagium

Throat

Chest

Undertail coverts

Wing panel

Primaries "hand"

Upperwing coverts

Secondaries

Nape

Rump

Superciliary line or "eye-line"

Uppertail coverts

Cheek

Back

Shoulder

Scapulars

Wing panel

FLIGHT POSITIONS

Soaring

Head-On

Gliding

Wing-On

Away

Stooping

Hovering

Flapping

Dihedral

Modified Dihedral

species, race, color, age, or sex, while others are shared among species. For example, the topside of the tail on a light Rough-legged Hawk is dark with a white base, but a dark Rough-legged Hawk lacks the white base on the tail. It is a good idea to understand traits thoroughly before relying on them in the field. Remember, it is more helpful to learn the basics of identification well before concentrating on the finer points. For example, birders should learn to recognize a Broad-winged Hawk's relatively stocky, pointed wings, and quick, choppy wing beats, with a secondary focus on plumage.

A good starting point to identifying a hawk to species is to decide what type of hawk (accipiter, buteo, falcon, etc.) it is. While this is often obvious based on structure, at times birds can change shape as they change postures. For example, Harriers have rounded wing tips when soaring, but their wings are strongly tapered and falcon-like when gliding. Also, birds can look and fly differently at ridge sites than over flat lands. For example, Ferruginous Hawks are often buoyant and wobbly along a ridge, but are fairly steady fliers over flat areas. Many birds are seen overhead or above eye level at flat land sites, but are often seen at or below eye level along a ridge. Therefore, experience at both types of sites is helpful.

Birders often attempt to identify raptors as quickly as possible, but studying birds with little concern for naming them "on the spot" aids in learning them. Snap identifications can be accurate only with experience and familiarity. It is also beneficial to watch a bird of known identity as it flies farther and farther away. Keep in mind that it is impossible to identify every bird; making accurate, positive identifications is much more rewarding than erroneously identifying every bird in the sky. **I have never met true experts who believe that they know everything regarding identification or can identify every bird they see.** Many accomplished hawk watchers are not the most vocal or authoritative persons in a crowd, but often the quiet ones off to the side.

HAWK COUNTING

Whether at a world-renowned migration site or fairly unknown one, counting hawks can be challenging and fun. If one's primary objective to counting hawks is to tally as many birds as possible, then there are certain techniques that will help in doing so. My approach is to focus primarily on birds passing nearby or overhead, and distant raptors secondarily. It requires more time to spot and identify distant birds than it does close birds. Therefore, the more time spent on distant birds, the greater number of birds that pass by undetected. Of course, distant birds should be monitored, but only after any nearby birds have been counted. Many distant birds eventually approach within sight as they pass. I have seen birders point out every "speck" on the horizon while failing to spot many of the nearby or overhead birds.

Probably the greatest factors influencing the number of hawks an observer counts are the effort of the observer and the scanning technique used. It is natural to scan across a horizon. However, **it can be much more effective to scan slowly in an up-and-down manner** (one field of view at a time) toward the origin of the main flight line. After scanning each side, scan across the sky overhead down to the horizon. When there are clouds overhead, it is productive to scan with the naked eye. Binoculars are often necessary with clear skies. **It is essential to scan slowly to spot high-altitude birds in a blue sky** that otherwise would be passed over with a quick scan. Resting binoculars on a tripod or armrest is a great way to stabilize them, making it much easier to focus on a distant subject.

OPTICS FOR HAWK WATCHING

There is no single binocular best suited for hawk watching. In a nutshell, most 7x (magnification) binoculars offer a wide field of view, while 10x binoculars offer added magnification. Regardless of the power, beginner birders should equip themselves with high-quality binoculars. Many birders (including myself) start off with an inexpensive model, only to buy a high-quality model later. Compact binoculars may offer a sharp image, but have a limited field of view, which makes locating birds difficult. There are several manufacturers that offer high-quality models. I have used Zeiss 7x45 Night Owls since they were introduced in 1994 and still love their sharpness and unmatched wide field of view. However, I acquired Zeiss Victory FL binoculars in 2008 and believe they offer the finest image available.

I highly recommend watching hawks with binoculars only! This is contrary to some birders who prefer to use high-powered spotting scopes. While using a scope, it is natural to focus on searching for plumage details of distant birds, and to disregard the shape and flight-style traits that a-re sometimes the only things visible with binoculars. Some observers will spot a bird in their binoculars and immediately try to locate it in their scope. Because of the scope's small field of view they never identify the bird as a result. Also, a spotting scope can be ineffective in gusty winds that cause it to shake (especially if it is mounted on an unstable tripod) or on days with strong heat shimmer that distorts an image. For those who prefer to use a scope for watching hawks, an eyepiece with a relatively large field of view is helpful. Fixed eyepieces with a single magnification tend to offer a wider field of view than zoom eyepieces (which offer various magnifications, but a small field of view). Zoom eyepieces are suitable for observing perched or idle birds, but fixed eyepieces are more suited for birds in flight. Overall, field experience is more valuable than high-powered optics in regard to identifying hawks.

PHOTOGRAPHY

Photographing birds in flight is exciting, but can be challenging. For raptor photography, a good auto-focus camera with a lens with a focal length between 300 and 500 millimeters is preferred because it has sufficient magnification and is easy to hand-hold and maneuver. Tele-converters add magnification to a lens, but may degrade the image quality slightly or slow the auto-focus and reduce the available light. Remember, the higher magnification the lens, the harder it is to hold the lens steady and locate birds in a viewfinder. A tripod will help stabilize a lens, but it can be awkward to use when tracking fast-moving birds or birds directly overhead.

Several manufacturers offer a variety of camera models that are excellent for action photography. Regardless of the brand, it is important to learn how to set up and use your camera. Many photographers keep their camera's factory default settings intact because they work well. Understanding how to use your camera's manual mode to select the proper settings for different conditions takes practice, but will optimize the chances of acquiring desirable photos. The manual mode allows for on-the-spot compensation in exposure with the turn of a dial. High ISO speeds will allow for faster shutter speeds; however, high ISO settings increase the "noise" (also known as "grain" when referring to film) in a photograph. All photographers improvise their techniques once they learn their subject, equipment, and surroundings.

Ethics

There are certain behavioral practices regarding wildlife photography that are unethical. The American Birding Association (ABA) reviews birding ethics in detail on their website. Some photographers overlook certain ethical issues, like trespassing on private property or altering habitat in any way, in order to capture a desired image. A rule of thumb is to emphasize the safety and welfare of a bird before one's personal agenda. Any exceptions to this rule are selfish and unethical. Photographing birds at nests is the foremost "no-no" regarding photography, and unfortunately the no-nest rule is being overlooked more often nowadays. Nest disturbance exposes eggs and chicks to predators and inclement weather, and may cause adults to abandon nest sites. While it is necessary for biologists to visit nest sites on occasion, photographers should never approach nests on their own.

Photographers should refrain from chasing or "bumping" birds for extended periods (more than a few minutes), and should never bump birds that are eating or holding prey. I have heard of photographers who have harassed a single hawk or owl from sunup to sundown. On top of this, they were using live bait in order to capture their desired images. Introducing a foreign food source into the wild is a risk to native wildlife. Even the most beautiful photograph becomes undesirable when found it was taken unethically. Photographers should enjoy witnessing a remarkable bird or event even if the chance to photograph it was unsuccessful.

Table 1

Timetable of Raptor Migration

Species	Spring	Fall
	March April May	Sept. Oct. Nov.
Black Vulture	••••••••●●●●●•••••••	•••••••••••••●●●●●●•••••••
Turkey Vulture	•••••••••••●●●●●••••••••	•••••••••••••••●●●●●••••••
Osprey	••••••••••••●●●●•••••	•••••••●●●●•••••••••••
Bald Eagle	••••●●●●●•••••••••••••	•••••●●●●●●••••••••••••
Mississippi Kite	•••••••••●●●•••••	•••●●●•••••••••••
Northern Harrier	••••••••••●●●●••••••••	•••••••••●●●●●●••••••●●•••••
Sharp-shinned Hawk	•••••••••●●●●●●•••••••	••••●●●●●●••●●●●•••••••••••
Cooper's Hawk	••••••••●●●●●••••••	•••••••●●●●●●••••••••
Northern Goshawk	•••••••••••●●●●•••••••••	•••••••••••●●●●●●••••••
Red-shouldered Hawk	•••••••●●●●••••••••••	•••••••••••••●●●●●•••••
Broad-winged Hawk	•••••●●●●•••••	•••••●●●●●●•••••••••••
Swainson's Hawk	••••••●●●●●••••••	••••••●●●●••••••••••
Red-tailed Hawk	••••••••●●●●●●••••••••	••••••••••●●●●●●••••••
Ferruginous Hawk	••••●●●●●••••	••••••••••●●●●●●••••••
Rough-legged Hawk	••••••••••••●●●●●••••	••••••••●●●●●●••••••••
Golden Eagle	•••●●●●●●••••••••	•••••••●●●●●●•••••
American Kestrel	••••••●●●●●●•••••	•••••••••●●●●●••••••••••••
Merlin	•••••••••●●●●••••	•••••••••●●●●●•••••••••••••
Peregrine Falcon	•••••••••••●●●•••••••	•••••••••••●●●●●•••••••••••
Prairie Falcon	••••••••••●●●••••••	•••••••••••●●●●•••••

Table 2

Occurrence of North American Raptors on Migration

Species	West	East
Black Vulture	Rare	Common except North
Turkey Vulture	Common	Common
Osprey	Fairly common	Common
Bald Eagle	Fairly common	Common
Mississippi Kite	Rare except Texas	Rare
Northern Harrier	Common	Common
Sharp-shinned Hawk	Common	Common
Cooper's Hawk	Common	Common
Northern Goshawk	Uncommon	Uncommon except Great Lakes region
Red-shouldered Hawk	Rare except California	Common
Broad-winged Hawk	Uncommon	Common
Swainson's Hawk	Common	Rare
Red-tailed Hawk	Common	Common
Ferruginous Hawk	Fairly common	Extremely rare
Rough-legged Hawk	Uncommon	Uncommon except Great Lakes region
Golden Eagle	Common	Fairly common
American Kestrel	Common	Common
Merlin	Common	Common
Peregrine Falcon	Fairly common	Common
Prairie Falcon	Fairly common	Extremely rare

Note: West = west of Minnesota and eastern Texas; East = east of North Dakota and western Texas.

SPECIES ACCOUNTS

ACCIPITERS

OVERVIEW

At first glance, **accipiters are recognizable in the field by their long-tailed, short-winged silhouette.** It is often easy to conclude that Sharp-shinned Hawks (the smallest accipiters) are small in size because of their quick movements and buoyant flight style, and that Goshawks (the largest accipiters) are large because of their slower, steadier flight. However, the three North American accipiters are very similar in appearance to each other in the field and judging size can be misleading. Distinguishing the accipiters from each other in flight requires considerable practice using a combination of shape, flight style, and plumage. Shape and flight-style traits are much more useful than plumage in identifying distant accipiters.

Among accipiters, **Sharp-shinned Hawks have stocky wings; the wings of Cooper's Hawks are slightly longer and slimmer; and Goshawks have broad-based wings with tapered "hands,"** making the wings quite angular along the back edge. Juvenile accipiters have slightly broader wings and longer tails than do adults, especially Goshawks. Sharp-shinned Hawks and Goshawks have stocky chests compared to those of Cooper's Hawks. **The head of Sharp-shinned Hawks is small and protrudes less beyond the wings than do those of Cooper's and Goshawks.** When soaring, the wings of accipiters are rounded at the tips, but when gliding or flapping, they become tapered, similar to those of falcons. Goshawks have the longest "hands" (primaries) of the accipiters and thus project farthest past the trailing edge of the wings when gliding; the wing tips of Sharp-shinned Hawks project the least.

The tail tip of all accipiters is rounded when spread. When folded, the tails of Sharp-shinned Hawks are square-tipped or rounded, those of Cooper's Hawks are rounded, and those of Goshawks' are rounded, wedge-shaped, or squared (some adults). The shape of the tail tip is not useful during spring due to wear or molt or in fall on actively molting adults (juveniles molt in spring through summer). Regardless of tail shape, **Cooper's and Goshawks have extremely long tails, whereas the tails of Sharp-shinned Hawks are shorter and slimmer.** The white tail tip of accipiters is typically most prominent on Cooper's Hawks. But this can be difficult to judge in the field, especially when backlit, as the white tail tip often appears unusually bright or wide.

Accipiters often fly directly, flapping and gliding intermittently more often than do other raptors, but with sufficient lift are proficient at soaring. **Sharp-shinned Hawks are extremely buoyant and often turbulent in flight,** but can be steady on light winds or when gliding. Cooper's and Goshawks are almost always steady fliers. While soaring, Sharp-shinned Hawks spiral in tight circles and rise more quickly than the lazy-soaring Cooper's and, especially, Goshawks. **The wing beats of Sharp-shinned Hawks are snappy and appear powerless; the wing beats of Cooper's Hawks and Goshawks appear stiff and forceful,** with Goshawks being quite deep at times. In flight, accipiters hold their wings flat or slightly drooped, but Cooper's Hawks often display a slight dihedral when soaring.

PLUMAGE

Juvenile accipiters are whitish below with dark streaking on the body. The amount of streaking varies on all three species, with Cooper's Hawks often showing the least. But this often is difficult to judge at a distance, so distinguishing accipiters on the basis of the underside plumage is often difficult. The topside of juveniles is brownish and indistinct; however, Goshawks often show paler upperwing coverts and heads than do Sharp-shinned and Cooper's Hawks.

Adult accipiters are blue-gray above and whitish below; adult Sharp-shinned and Cooper's Hawks have pale rufous barring on the body, while Goshawks have faint black barring. Adult male accipiters are bluer than females on top with slightly darker "hands." Juveniles and adult females typically show little contrast above; even in low light, ageing or sexing of some birds may be possible. The two-toned upperwings of adult male accipiters can be helpful when telling adult male Cooper's Hawks from female Sharp-shinned Hawks, which are similar in size and shape. Male accipiters are considerably smaller than females, often exhibiting quicker wing beats and more tapered wings. Thus, with practice it is possible to sex them by shape and flight style. All accipiters acquire their adult plumage after their first molt starting at about one year old. Many adults retain some juvenile feathers throughout their second year, but this is impossible to see at a distance, except on some Goshawks.

Sharp-shinned Hawk

Sharp-shinned Hawk

(Accipiter striatus)

OVERVIEW

Sharp-shinned Hawks, or "sharpies" as they are commonly called, are the smallest North American raptor (about the size of a jay) and the most common and widespread of the accipiters. **They are stocky overall, with short, broad wings, long, narrow tails, and small heads.** While the head of a Sharp-shinned Hawk projects slightly past the wings when soaring or gliding, it does so visibly less than on Cooper's Hawk and Northern Goshawk. Sharp-shinned Hawks with full crops display an exaggerated head projection, making them appear more similar to Cooper's Hawks. Female Sharp-shinned Hawks are slightly larger than males, and sexes are nearly identical in plumage.

Sharp-shinned Hawks are small, buoyant, and unsteady, only appearing steady when flapping into a headwind or soaring on light winds. **The wing beats of Sharp-shinned Hawks are very quick and lack power.** When flapping vigorously in chase of prey or into a headwind, Sharp-shinned Hawks exhibit pointed wings and resemble a small falcon. **Of the accipiters, female Sharp-shinned Hawks and male Cooper's Hawks appear the most similar in size and shape,** and distinguishing the two often is difficult. On the contrary, there are obvious differences between female Cooper's Hawks, which are large and lengthy overall, and male Sharp-shinned Hawks, which are diminutive and compact.

PLUMAGE

Juveniles **are whitish below with dark streaking on the body and brown on the upperside.** Many juvenile Sharp-shinned Hawks have dense, rufous streaking underneath that appears similar to the barred underside of adults, and these barred juveniles can be particularly difficult to age in the field. Juvenile Sharp-shinned Hawks may also look orangey underneath when the sun is low on the horizon. All accipiters have banded tails, but the banding is indistinct and not helpful in flight identification. However, **the white tail tip that all accipiters show is typically narrowest on Sharp-shinned Hawks.**

Adults **are pale below with rufous barring and blue-gray on top.** Adult males are slightly more colorful than are adult females, especially on the wing pits. Adult males have blue upperwing coverts that contrast with darker remiges compared to the more uniform upperwing or sometimes paler-"handed" females. This is helpful when trying to tell adult female Sharp-shinned Hawks from adult male Cooper's Hawks. Plumage of accipiters fades by spring, and adult females can appear brownish on top similar to juveniles whereas adult males can look pale blue on top. In contrast, juveniles in autumn can appear to have a grayish sheen on the topside and look similar to adults.

SS 01 - **Juvenile Sharp-shinned Hawks** are pale below with dark streaking on the chest (top left), but appear whitish in direct sunlight (top right). The **juvenile** streaking on the body is difficult to observe when shadowed or backlit (middle). Many **juveniles** show rufous streaking on the body that looks like the rufous underside of adults, especially at a distance (bottom left) or when flapping at eye level (bottom right). All Sharp-shinned Hawks are stocky with small heads and narrow, long tails.

SS 02 - Near sundown, the underside of **juveniles** can appear rufous like that of adults (top left). In poor light, Sharp-shinned Hawks can appear completely dark and be difficult to age (top right). In strong winds along a ridge, Sharp-shinned Hawks may fly with wings fully tucked (middle left) or cock their tails up (middle right). Gliding head-on, Sharp-shinned Hawks are stocky overall and hunched at the shoulders. **Juveniles** are brown on the head (bottom left); **adults** are blackish-gray on the head and rufous on the chest (bottom right).

SS 03 - **Adults** in good light (top left) or direct sunlight (top right) show a rufous underside and white undertail coverts. Against a white sky, the rufous underside of **adults** is often obvious (middle left). When **adults** are flapping at eye level, their wing pits appear vibrant, especially those of males (middle right). Plumage of **adults** can be difficult to see when shadowed (bottom left) or backlit (bottom right).

SS 04 - Near sundown, the rufous underbody of **adults** is particularly orange (top left). When skies are overcast, the plumage of **adults** when seen head-on may be obscured (top right). Telling Sharp-shinned from Cooper's Hawks at certain angles can be impossible, but note the short wings and tail and stocky body of the Sharp-shinned Hawk (bottom).

SS 05 - **Juveniles** are brownish on top and indistinct (top left). Some **juveniles** appear grayish-brown on top (top right); note the tawny head like that of juvenile Cooper's Hawks. In spring, **juveniles** can show bluish backs as they begin to acquire adult plumage (middle left). **Adult males** are blue on top with darker "hands" (middle right); **adult females** are often more uniform in tone (bottom left). In fall, only **adults** can show flight feather molt (bottom right).

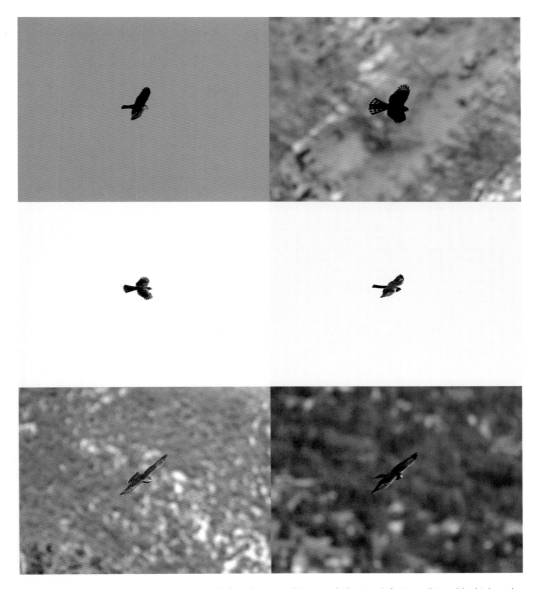

SS 06 - Sharp-shinned Hawks can appear darker than usual in poor light (top left, juvenile) or blackish and impossible to age (top right). When side-lit, **juveniles** (middle left) and **adults** (middle right) can appear dark overall on top, but show color at times. In spring, the topsides of **juveniles** (bottom left) and **adults** (bottom right) are often faded and appear similar to each other.

Cooper's Hawk

Cooper's Hawk

(Accipiter cooperii)

OVERVIEW

Cooper's Hawks are larger than Sharp-shinned Hawks but almost identical in plumage and very similar in shape. Cooper's Hawks from the West are smaller on average than in the East, and appear even more like Sharp-shinned Hawks, especially male Cooper's Hawks. However, Cooper's Hawks always display relatively longer wings and tails and larger heads. As well, they are steady fliers with somewhat stiffer, less furious, powerful wing beats compared with Sharp-shinned Hawks. They soar on flat wings or with a slight dihedral, and glide on slightly bowed wings. Females are larger than males and generally appear sizeable in the field, but males often seem no larger than Sharp-shinned Hawks.

In a glide, Cooper's Hawks look compact, similar to Sharp-shinned Hawks, but their heads and tails extend farther, and they show longer, less squared "hands" in comparison. Cooper's Hawks in active molt can exhibit square-tipped tails or squared-off wings, making them appear even more similar to Sharp-shinned Hawks. Also, the farther away a Cooper's Hawk is, the more similar it appears to a Sharp-shinned Hawk.

PLUMAGE

Juvenile Cooper's Hawks are whitish below with dark streaking on the body and dark brown above. Cooper's Hawks typically show the least prominent streaking of the accipiters, but the streaking on the underside varies in density. The underside plumage is often not useful in identifying distant juvenile accipiters. Juvenile Cooper's Hawks show a tawny wash to the head. While juvenile Goshawks may show this trait, most Sharp-shinned Hawks do not. As with Sharp-shinned Hawks, Cooper's Hawks have indistinct banded tails; however, the white tail tip is usually more prominent on Cooper's Hawks than on Sharp-shinned Hawks. When backlit against a blue sky, the white tip to the tail of Sharp-shinned Hawks can appear quite prominent.

Adult Cooper's Hawks are pale below with rufous barring and blue-gray on top. Adults have a dark cap and a pale nape, whereas adult Sharp-shinned Hawks lack the pale nape and thus a distinct cap. Male Cooper's are slightly more colorful than females, especially on top, where they show blue upperwing coverts and dark primaries compared to the more uniform gray-blue upperwing of females. Some adult females are a brownish-gray above, similar in color to that of juveniles, especially in spring if they are heavily faded. Underside plumage of Cooper's Hawks can fade by spring and appear whitish. Adult male Cooper's Hawks have gray cheeks, but males in their first year of adulthood often have rufous cheeks like most females.

CH 01 - Juvenile Cooper's Hawks are pale below with brown streaking on the chest (top left). In direct sunlight, **juveniles** (shown, top right) and adults can appear all white below. When shadowed (middle left) or in poor light (middle right), Cooper's Hawks can appear overall dark. When backlit, plumage and wing shape can be difficult to judge (bottom left, juvenile). **Juveniles** show tawny heads when seen head-on (bottom right). Note bowed wings of all Cooper's Hawks when gliding. Note the lengthy overall shape compared to that of Sharp-shinned Hawks.

CH 02 - The rufous underside of **adults** is obvious in good light (top left) or over snow (top right). When backlit (middle left) or shadowed (middle right), the plumage of **adults** can be difficult to discern. In fall, **adults** may show flight feather molt, causing squared wing tips and tail tip (bottom left). Head-on, the pale cheeks and dark cap of **adults** are sometimes obvious (bottom right). Adult Cooper's Hawks are overall stockier than juveniles and more similar in shape to Sharp-shinned Hawks.

CH 03 - **Juveniles** are brown on top with a tawny head (top left). **Adult males** are blue on top with dark primaries, gray cheeks, and a dark cap (top right). **Adult females** are uniform blue-gray on top with pale cheeks and a dark cap (middle left). Molting or faded **adult males** can appear brownish on top and square-tailed; note the gray cheeks but lack of a white tail tip (middle right). Some **adult females** are normally brownish-gray on top; note the dark cap and flight feather molt (bottom left). In poor light, **adults** (shown) and sometimes juveniles can appear blackish on top (bottom right).

CH 04 - Many **Sharp-shinned Hawks** have rounded tail tips when closed (top left); **Cooper's Hawks** in spring often show squared tail tips when closed due to wear (top right). Sharp-shinned Hawks with full crops (middle left) can appear large-headed and similar to Cooper's Hawks (middle right). When backlit, the white tail tip of both Sharp-shinned Hawks (bottom left) and Cooper's Hawks (bottom right) can appear prominent. Note the larger head and longer wings and tail of the Cooper's Hawk compared to the Sharp-shinned Hawk.

Northern Goshawk

Northern Goshawk
(Accipiter gentilis)

OVERVIEW

Northern Goshawks are the largest of the North American accipiters and the least common at most migration sites. Their wing shape often is the telling feature separating them from Cooper's and Sharp-shinned Hawks. **The wings of Goshawks are broad and taper toward the tips, appearing stocky in a soar and falcon-like in a glide.** The tail of Goshawks is very long and fairly broad, and the chest is bulky. Males, which are smaller than females, have slimmer, more tapered wings, especially adults, but this can be difficult to notice without considerable practice. In flight, **Goshawks are steady and powerful with stiff, labored wing beats, which are loftier on the upstroke than those of the similar Cooper's Hawk.** Males can exhibit quick wing beats that are tricky to recognize as a Goshawk's. Goshawks hold their wings flat or slightly drooped at all times. Cooper's Hawks (and sometimes even Sharp-shinned Hawks) can be perceived as large in the field, and are often misidentified as Goshawks, but judging accipiters by size alone often is inaccurate.

PLUMAGE

Juvenile Goshawks are pale below with dark streaking on the underside. **Most juveniles are heavily streaked underneath,** but some individuals are lightly streaked and appear pale overall. Goshawks typically have streaking on the undertail coverts, but this is often a useless field mark on birds that are distant or shadowed. Juveniles are brown on top with slate and tan highlights throughout that sometimes give Goshawks a more grayish appearance on top than Sharp-shinned and Cooper's Hawks. **Goshawks typically show pale mottling along the upperwing coverts that forms a narrow bar,** but some birds (especially in the West) lack this trait. Some Cooper's Hawks may show mottling or fading (in spring) on the upperwings, but it is rarely as prominent as on Goshawks. **Juvenile Goshawks have pale cheeks and a broad, pale superciliary line, which can make them look pale-headed in the field,** distinguishing them from Sharp-shinned and Cooper's Hawks.

Adult Northern Goshawks are pale grayish-white below with faint, dark barring on the body, **appearing pale overall at a distance.** Sometimes the flight feathers are slightly darker in contrast, making them appear two-toned underneath. In bright sunlight, the head and body can appear gleaming white. Adult males are bluish on top with dark primaries, or "hands," and slightly more colorful than adult females, which are more uniformly gray-blue on top. Rarely, females are slate-brown on top, appearing similar to juveniles. The head of adult Goshawks is blackish with a prominent white superciliary line. In direct sunlight from a head-on perspective, the bluish uppersides of Goshawks can appear almost whitish, especially in males in spring.

NG 01 - Juvenile Goshawks are buffy below with prominent dark streaking on the body (top left). When backlit, the dark streaking of **juveniles** can be difficult to see (top right). In direct sunlight, **juveniles** often look whitish underneath (middle left) or pale-headed at eye level (middle right). In spring, many **juveniles** appear whitish underneath due to fading (bottom left). Gliding head-on, Goshawks show broad chests and drooped wings (bottom right, juvenile). All Goshawks have long tails and stocky wings that taper toward the tips.

NG 02 - Adult Goshawks are pale gray below, appearing overall unmarked (top left). In good light, the underwings of **adults** appear two-toned (top right). In direct sunlight, **adults** appear white below (middle left), especially when flapping (middle right). Head-on, **adults** show pale chests and blackish heads with pale cheeks and a white superciliary line (bottom left). When backlit, the flight feathers of **adults** appear pale; note the uniform gray underside (bottom right). Note the short, broad, tapered wings and long tail that all **adults** show.

NG 03 - Over snow, **juveniles** (top left) and **adults** (top right) appear unusually bright. When shadowed, **juveniles** can appear uniformly dark underneath, and the streaking on the body may be difficult to see (middle left, right). **Adults** can appear dark with paler flight feathers when backlit (bottom left). When shadowed, **adults** may appear uniformly dark, but the pale cheeks are still obvious in direct sunlight (bottom right).

NG 04 - Juveniles are brownish on top with a pale, narrow, mottled upperwing bar (top left). Some **juveniles**, particularly in the West, lack mottling on the upperwings (top right). In spring, many **juveniles** show faded upperwings (middle left). **Adult male** Goshawks are blue on top with blackish flight feathers (middle right); **adult females** are more uniformly blue-gray on top (bottom left). The upperside of **adults** can look blackish in poor light (bottom right).

Northern Harrier

NORTHERN HARRIER

(Circus cyaneus)

OVERVIEW

The Northern Harrier is the only North American representative of the genus *Circus*. Harriers are often compared to owls, due to their prominent facial disk, acute hearing, and ability to hunt and migrate in darkness. However, they hunt most often during the day, coursing low over fields and marshes; hence their former name "Marsh Hawk." **Harriers are distinctive in flight,** often identifiable by their delicate side-to-side teetering mannerism, especially at low altitudes when searching out prey. Although wobbly in demeanor, Harriers are able to maneuver quickly and display bursts of speed like other hawks. **Harriers are extremely buoyant,** appearing weightless as they gain lift with great ease. They nearly always **hold their wings in a prominent dihedral or modified dihedral.** The only time Harriers fly with drooped wings is when gliding on strong ridge updrafts, when they can also be quite steady. Unlike many raptors, Northern Harriers traverse large bodies of water or desert during migration utilizing **lofty, relaxed, fluid wing beats** that are even in tempo and lack stiffness.

Harriers are slim overall with long, slender wings and long, square-tipped tails. In a glide, they appear Peregrine-like, except their wings and tail are less tapered and they lack a broad head and chest. The wings of Harriers do not bulge along the secondaries like those of buteos, and their heads are relatively small. Male Harriers have proportionately shorter wings and tails than females, and often display a slightly shallower dihedral. With practice, separating males from females based on shape and flight style is possible.

PLUMAGE

Adult female and juvenile Northern Harriers are **dark brown on top and pale below with dark flight feathers.** Juveniles have rufous underparts with faintly streaked chests. Adult females are whitish to rufous underneath with dark streaking on the body; some adult females show limited streaking, which is invisible at a distance similar to juveniles. In spring, the undersides of adult females and juveniles fade to whitish, causing them to look identical to each other at a distance. Adult female and juvenile Harriers are almost identical on top. Juveniles are uniform in tone; adult females often show a golden nape and a grayish tone on the flight feathers, creating a slight contrast with the plain brown upperwings. All Harriers have bright white uppertail coverts, often referred to as a white "rump."

Adult male Northern Harriers are **snow-white underneath with black wing tips and a black trailing edge to the wings,** which juveniles and adult females do not show. Some males, typically birds in their first adult plumage, have orangey barring to the chest and undertail coverts. **The upperside of adult males is grayish or grayish-brown,** which can fade to pale gray by spring. All Northern Harriers acquire their adult plumage after their first molt at about one year old.

There are three records of dark Northern Harrier in North America. Dark birds are completely blackish on the underside with paler flight feathers and blackish on top, lacking white uppertail coverts.

NH 01 - Juvenile Northern Harriers are rufous underneath with dark flight feathers, and lack visible streaking on the body (top left, right). **Adult females** are buffy (middle left) or rufous (middle right) underneath with dark streaking, but can be impossible to tell from juveniles at a distance or at eye level (bottom left). At sundown, **adult females** (bottom right) can appear particularly rufous and look very similar to juveniles.

NH 02 - Telling juvenile from adult female Harriers when backlit (juvenile, top left) or in poor light (juvenile, top right) can be impossible. Plumage of Harriers can be difficult to discern at dusk (adult female, middle left) or when shadowed (adult female, middle right). Adult females and juveniles fade to buff underneath by spring; note that juveniles still lack streaking (bottom left). Dark Harriers are extremely rare but do occur; note the dark underbody with paler flight feathers and lack of white "rump" (bottom right).

NH 03 - Adult male Harriers are white below with black wing tips and a black trailing edge on the secondaries (top left). Some **adult males** have a rusty-brown chest and undertail coverts, which can be difficult to discern (top right). Rarely, males in their first adult plumage retain juvenile outer primaries and lack solid black wing tips (middle left). When backlit, **adult male** Harriers appear to have pale primary panels (middle right), but when shadowed can appear dark underneath (bottom left). At sundown, **adult males** can appear rufous underneath (bottom right); note the black wing tips.

RS 02 - Adult Red-shouldered Hawks are rufous on the underbody with dark flight feathers showing a dark trailing edge and black-and-white banded tails (top left). The pale wing commas on **adults** are obvious when backlit (top right), but can be impossible to see when shadowed (middle left) or in a glide (middle right). Head-on, **adults** show rufous chests and pale faces; note the drooped, squared wings (bottom left). In poor light, **adults** can appear dark underneath, but the wing commas may be visible (bottom right); note the somewhat long tail.

RS 03 - California juveniles are marked rufous-brown underneath similar to adults, but are more heavily marked on the chest and lack a dark trailing edge on the wings (top left, right). California **adults** are uniform pale rufous underneath with a dark trailing edge on the wings (middle left); the black-and-white banded tail of California adults and juveniles is similar. **Eastern juveniles** are brown on top with buffy wing commas (middle right); California **juveniles** have blackish flight feathers, and white wing commas (bottom left). **Adults** of all races have boldly banded wings and tails and rufous shoulders (bottom right).

Broad-winged Hawk

(Buteo platypterus)

OVERVIEW

Broad-winged Hawks are small, chunky buteos with **stocky, pointed wings** and a bulky head and chest. The tail is narrow and usually square tipped, appearing accipiter-like when closed. From a side view, Broad-winged Hawks resemble accipiters, but their heads and chests are bulkier, their wings are broader and sharply pointed, and they have shorter tails. When Broad-winged Hawks are gliding overhead, their wings appear particularly stocky, and the wing tips barely project past the back edge of the wings. Broad-winged Hawks hold their wings flat or with a slight droop, and are steady fliers except in the most extreme winds; they soar in tight circles compared with larger buteos. Broad-winged Hawks flap in a quick, shallow manner similar to Red-shouldered or Cooper's Hawks, but their wing beats are stiff or choppy in comparison. Broad-winged Hawks often can be seen in large flocks, or kettles, on migration, especially in the East and along the Texas coast.

PLUMAGE

Juvenile light Broad-winged Hawks are pale underneath with dark streaking that is often limited to the sides of the chest but can vary from almost nonexistent to prominent throughout the body. Some juveniles even show a bellyband similar to that of Red-tailed Hawks (but lack other Red-tailed Hawk traits like dark patagials). Rarely, juvenile Broad-winged Hawks are somewhat barred on the chest similar to adults. **Overall, juveniles appear pale underneath at a distance.** The underwing coverts of juveniles are essentially unmarked, and the underside of the tail is pale with a dark sub-terminal band, which is the only band visible in the field. A few juveniles can show a boldly banded tail that appears identical to those of adults, but they lack other adult traits, like a dark trailing edge to the wings or rufous underbody. Males and females are identical in plumage.

Juvenile dark birds are uniformly dark on the body and underwing coverts; some dark birds show sparse white mottling underneath. The flight feathers are identical to those of light birds. The topside of all juveniles is brown with sparse to moderate pale mottling along the upperwings and pale primary panels (similar to those of Red-tailed Hawks but less distinct). Dark juveniles can be slightly darker overall and show less mottling on top than do light birds, but this is difficult to judge.

Adult light Broad-winged Hawks are pale underneath with rufous barring on the chest (and sometimes on the belly). The barring on some adults is faint and inconspicuous in the field, while others show a solid rufous chest. A few Broad-winged Hawks in their first year of adulthood show juvenile-like body plumage. All adults have a defined dark trailing edge to the wings, whereas juveniles rarely do. **The tail of adults from underneath is black with a bold white band toward the base.** All adult Broad-winged Hawks are dark brown on top with slightly darker flight feathers. The topside of the tail is black with a bold white band at the center; many show a narrow white band at the base, but this is difficult to see at a distance.

Broad-winged Hawk

When seen head-on, adults appear dark-headed, whereas juveniles are often paler on the crown or face.

Adult dark Broad-winged Hawks are completely blackish on the body and underwing coverts. The banded tail of dark birds is identical to that of light birds. Dark birds are uncommon, especially in the East. Be careful, as birds in poor light can appear uniformly dark, but true dark birds are blackish underneath with paler flight feathers.

BW 01 - Juvenile Broad-winged Hawks are pale underneath with dark streaking on the body, appearing nondescript; note the pale wing panels (top left). When shadowed, the stocky, pointed wings are the most identifiable trait (top right). Some **juveniles** are unmarked underneath (middle left) or show a bellyband (middle right); note the dark tip on the fanned tail. In spring, molted primaries (bottom left) appear similar to the wing commas on Red-shouldered Hawk. Head-on, **juveniles** are brownish with paler heads; note the stocky appearance (bottom right).

BW 02 - A few **juveniles** have adult-like tails (top left) or a dark trailing edge to the wings (top right) but lack adult body plumage. Some **juveniles** are heavily barred on the underbody and look similar to adults (middle left). Wing-on, Broad-winged Hawks are similar to accipiters, but have larger heads and shorter tails; note the pale primary panels on **juveniles** (middle right). **Dark juveniles** are dark underneath with pale flight feathers; note the dark tail tip (bottom left). **Dark adults** have a banded tail and a dark trailing edge to the wings (bottom right).

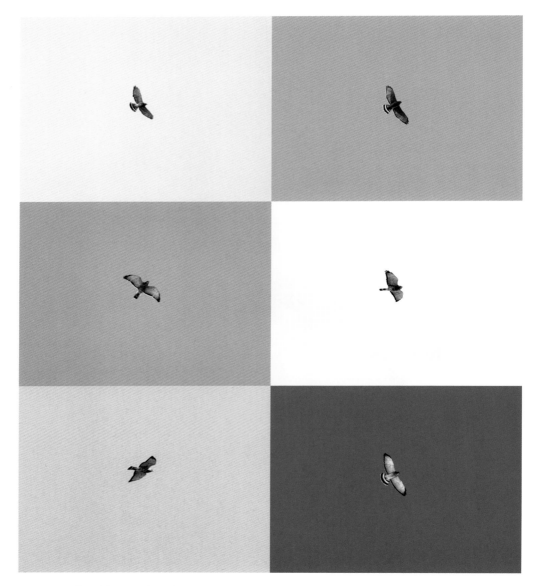

BW 03 - **Light adults** are pale underneath with rufous chests and boldly banded tails (top left). When shadowed, the bands are still obvious on a fanned tail (top right). When gliding, tail bands may be difficult to see, but note the dark trailing edge to the wings and the rufous chest in good light (middle left) and when backlit (middle right). Headed away, Broad-winged Hawks (bottom left) appear similar to accipiters, but are overall stockier with shorter tails; note the dark trailing edge on wings. Over snow, plumage of **adults** appears more prominently (bottom right).

BW 04 - Head-on, **adults** (top left) appear darker overall than juveniles. The indistinct topside of **juveniles** is brown with pale mottling along upperwings (top right). Wing-on, **juvenile** Broad-winged Hawks (middle left) appear accipiter-like in shape and plumage, but note the shorter tail with a dark tip. Molting **juveniles** in spring can appear to have wing commas like Red-shouldered Hawks but lacking a border (middle right). **Adult** topside is dark brown with an obvious black-and-white banded tail (bottom left). At eye level, tail bands of **adult** can be difficult to see (bottom right).

Swainson's Hawk

Swainson's Hawk

(Buteo swainsoni)

OVERVIEW

Swainson's Hawks are large buteos with long, somewhat narrow wings that **are pointed at all times.** Their wing shape is similar to that of a Broad-winged Hawk, but is a lengthy, more elegant version. When Swainson's Hawks are gliding at eye level, their wings appear Osprey-like but shorter and less drooped. When Swainson's Hawks are **gliding overhead, their wing tips project well past the back edge of the wings, creating a "M"-shaped silhouette.** Swainson's Hawks soar with a pronounced dihedral and are buoyant or wobbly but much less so than Harriers. **The wing beats of Swainson's Hawks are somewhat deep and fluid,** but adults usually show shallower, stiffer wing beats than do juveniles. When coursing low over a field, they are sometimes mistaken for Northern Harriers, but they have stiffer wing beats and a broad chest in comparison. Swainson's Hawks can be seen in large kettles on migration, especially in the West and along the Texas coast.

PLUMAGE

Light **Swainson's Hawks are whitish underneath with dark remiges,** appearing two-toned, or "black in back." The flight feathers of juveniles are somewhat paler underneath than those of adults, making them less distinctly two-toned. *Juveniles* show a pale rufous body that fades to whitish by winter, and dark streaking that can be either limited to the sides of the chest or marked throughout the body. Juveniles can have dark or pale heads in fall, but the head is almost always pale by spring due to fading; adults always show dark heads. *Adult* **light Swainson's Hawks have a dark chest bib** that is apparent at considerable distances. In some cases, juveniles are heavily streaked on the chest and appear bibbed. The tail of all Swainson's Hawks appears somewhat pale underneath with a **dark tip that is broader on adults than on juveniles.** Swainson's Hawks lack the pale primary panels most other juvenile buteos show.

From above, the **blackish flight feathers of Swainson's Hawks contrast with the brown upperwing coverts, appearing two-toned** (or often dark-"handed"), especially on juveniles, which have pale-fringed upperwing coverts. In spring, these pale fringes are often absent due to wear. However, the two-toned upperside of all Swainson's Hawks is more pronounced in spring, when the upperwing coverts are quite pale from fading. All light Swainson's Hawks have pale uppertail coverts that contrast with the tail and "rump," but the uppertail coverts are not nearly as distinct as on Northern Harriers. At eye level or in poor light, Swainson's Hawks (especially adults) can appear blackish or raven-like on top. Swainson's Hawks can appear uniformly dark underneath when shadowed, but when backlit can appear pale overall (especially juveniles) or even dark with pale flight feathers. There is overlap in plumage between male and female Swainson's Hawks and determining sexes in the field can be inaccurate. Plumage of juvenile males and females is identical in the field.

Dark **Swainson's Hawks are uniformly dark underneath.** *Adults* appear solid dark on the body, but usually show slightly paler underwing coverts. *Juveniles* are moderately to heavily streaked on the underbody and sometimes on the underwing coverts, but typically appear dark underneath with paler underwing coverts. The topsides of dark Swainson's Hawks are similar to those of light birds, but are often a shade darker overall and may show dark uppertail coverts. There is a continuum of plumages from light to dark in Swainson's Hawks, and some birds are difficult to categorize. Birds with plumages in between typical light and dark birds appear more similar to dark birds from a distance.

Swainson's Hawks acquire a *sub-adult* plumage at about one year old. Sub-adults are juvenile-like on the underbody with adult-like flight feathers (remiges are darker than those of juveniles with a broad, dark trailing edge), and are adult-like on the topside. **Sub-adults often have pale or whitish heads in spring similar to those of juveniles.** In autumn, Swainson's Hawks that show symmetrical wing or tail molt, but exhibit juvenile body plumage, are sub-adults. A fair number of sub-adults have an adult-like body plumage and appear identical to adults in the field.

SW 01 - **Light juvenile Swainson's Hawks** are pale underneath with dark flight feathers and streaking on the sides of the chest (top left) and belly (top right). **Dark juveniles** are streaked throughout the underside with dark flight feathers (middle left), appearing uniformly dark, especially when shadowed (middle right). Some heavily marked **juveniles** have a dark bib and appear adult-like (bottom left). When shadowed, **light juveniles** can appear dark underneath (bottom right). Note the long, pointed wings characteristic of all Swainson's Hawks.

SW 02 - **Light juveniles** in spring show pale heads and whitish underbodies (top left). When backlit, the **juvenile** flight feathers can appear pale and the underbody can appear dark (top right). **Light sub-adults** show adult flight feathers but lack a bib (middle left); however, some appear bibbed and can be impossible to age (middle right). **Sub-adults** in spring (dark bird, bottom left) have pale heads like that of juveniles; note the streaked body. **Light adults** have a full bib on the chest (bottom right).

SW 03 - **Light adults** show less two-toned wings than usual when backlit (top left) or shadowed (top right); note the complete bib, but uniform dark appearance in poor light (middle left). Head-on, the dark head and bib of **light adults** are often visible; note the long, bowed wings (middle right). **Light adults** can have varying amounts of barring on the belly, but still appear pale on the belly and underwing coverts in good light (bottom left) or when shadowed (bottom right).

SW 04 - Some **light adults** are nearly solid dark on the upper body with pale lower bellies; note the blunt wing tips due to molt (top left). **Dark adults** are solid dark on the body with paler underwing coverts (top right), or completely dark underneath (middle left). When shadowed, **dark adults** may appear uniformly blackish underneath (middle right). Most **juveniles** show pale mottling along the upperwings (bottom left), but some may not (bottom right).

SW 05 - **Juveniles** in spring have pale heads and lack mottling to the upperwings (top left); they can appear identical to **sub-adults** (top right) at a distance. **Light adults** are somewhat more two-toned than juveniles and show dark heads year-round (middle left). **Dark adults** are darker overall and less two-toned above than light birds (middle right). **Adults** in spring are paler on top than usual due to fading (bottom left). Head-on, **adults** can appear almost blackish above (bottom right). Note the two-toned upperwings characteristic of all Swainson's Hawks.

Red-tailed Hawk

Red-tailed Hawk

(Buteo jamaicensis)

OVERVIEW

Red-tailed Hawks are large, stout, and powerful—the quintessential North American buteo. Red-tailed Hawks are able to hover like Rough-legged Hawks, soar like Swainson's Hawks, and stoop like Golden Eagles. They have **long, broad wings and broad tails, and appear steady in flight,** soaring in wide circles on flat wings or with a slight dihedral. They glide on slightly bowed wings, appearing particularly stocky when doing so. **In a shallow glide, the wing tips protrude slightly past the trailing edge of the wings,** but project well past when in a steep glide. The wing beats of Red-tailed Hawks appear shallow and powerful but fluid, but can be deep and labored when they are accelerating forward. Juveniles have slimmer wings and longer tails than do adults, and show a more exaggerated dihedral. Juveniles will often exhibit fairly floppy wing beats as opposed to the choppier wing beats of adults.

There are several races of the Red-tailed Hawk, but the Eastern (*B.j. borealis*), Western (*B.j. calurus*), and Harlan's (*B.j. harlani*) are migratory. All other races in North America are largely non-migratory and indistinct in plumage from Eastern or Western in the field. Eastern birds show a light plumage only; Western and Harlan's races occur in a continuum of light and dark plumages. The ranges of these three races meet in specific areas where interbreeding among the races likely occurs. However, likely intergrades between these races are nearly impossible to identify in flight. The Eastern race is seen throughout much of North America on migration, being uncommon in Alaska and west of the Rocky Mountains. Western and Harlan's are seen throughout western North America and the Midwest, reaching the Gulf Coast states in fall and winter, but Western birds are uncommon in Alaska.

Red-tailed Hawks can vary greatly in shape. Smaller birds are typically quite stocky and larger birds quite lanky. Western Red-tailed Hawks are generally longer-winged than Eastern, with Harlan's race being the most variable in shape. Regardless, races are impossible to discern by wing shape alone. Red-tailed Hawks can sometimes be seen in small kettles on migration.

PLUMAGE

Red-tailed Hawks are highly variable in plumage, ranging from completely whitish to completely blackish on the body. *Light* **Red-tailed Hawks are brown above and whitish below with dark patagial bars and with most showing dark bellybands.** Excluding most Harlan's, adults have a reddish tail. *Juveniles* have brownish, indistinct tails, but some juveniles, especially Eastern birds, have unusually reddish tails. From below when folded, the tail looks pale and unmarked on both adults and juveniles. *Adult* **Red-tailed Hawks have a bold, dark trailing edge to wings, but lack the pale primary panels of juveniles.** At certain times, adults can appear to have dark flight feathers like those of Swainson's Hawks, but they are not two-toned underneath to the extent of Swainson's Hawks. Adults often show a pale rufous wash to the underside compared to the whitish tone of juveniles. In early fall, juveniles may have a strong buffy wash to the breast before fading to whitish. In spring, juveniles of all races can have unusually pale heads due to fading.

Dark Red-tailed Hawks are dark brown underneath with pale flight feathers. Most *juvenile dark* birds are heavily streaked underneath, but some are uniformly dark on the body. The underwing coverts are dark throughout, masking the distinct patagial bars that light birds show. *Adults* can be solid dark underneath, but **most dark Western adults have dark rufous chests and most dark Harlan's adults have white mottling on the chest.** Dark Red-tailed Hawks are slightly darker overall, sometimes blackish, on top than are light birds.

Eastern

Eastern Red-tailed Hawks typically show light to moderately marked bellybands, unmarked underwing coverts, and white throats. Eastern birds breeding in Canada and the northernmost United States may be very heavily marked underneath and show dark throats similar to those of heavily marked Western birds. On the other hand, some Eastern birds, especially the *Krider's* form that breeds primarily in the Dakotas, eastern Montana, Manitoba, and Saskatchewan, can be virtually unmarked underneath. **Krider's are unique in that they often have whitish heads and tails and extensive white mottling on the upperwings.** Krider's has yet to be documented west of the Rocky Mountains, so a Red-tailed Hawk west of the Rockies with a whitish tail is likely a Harlan's race.

Western

Light Western Red-tailed Hawks typically show moderate bellybands, dark mottling to the underwing coverts, and broad patagial bars. Western birds with lightly marked bellies and underwing coverts can still exhibit broad patagial bars and dark throats, whereas lightly marked Eastern birds often show light throats and faint patagial bars. Western birds do not approach the pale extreme of Easterns or show a whitish tail or head. **Many *Western* adults have a prominent rufous wash to the underside** that most Eastern birds do not show. Like adults, *juvenile Western* Red-tailed Hawks are generally more extensively marked underneath than Eastern birds. Regardless, **there are many Western birds that are inseparable from Eastern birds in the field.**

Adult dark Western **Red-tailed Hawks are dark brown on the belly and underwing coverts with a dark rufous-brown chest that contrasts slightly with the belly.** At a distance or when shadowed, the two plumage types are often impossible to differentiate. *Dark Western juveniles* are heavily streaked underneath, but some are solid brown on the body. Most streaked birds have slightly less prominent streaking on the chest, looking somewhat pale-chested in bright light, but all types can appear uniformly dark at a distance.

Harlan's

The tail pattern on Harlan's adults often is markedly different than that of other races. Light Harlan's adults have grayish, brownish, or whitish tails often with a broad, blackish or rufous tip. From below, the tail of Harlan's appears whitish similar to that of other races. Some Harlan's adults show an almost completely reddish tail with limited whitish, gray, black, or brown mottling, making them difficult to discern from other races. Dark Harlan's adults usually have grayish or whitish tails with a dark tip. A few dark Harlan's adults have an overall blackish tail or a dark tail with faint banding. The tail of Harlan's juveniles is typically inseparable from those of other races at a distance.

Light Harlan's juveniles and adults **are snow-white underneath**, with only some showing the buffy undersides of other races. Be aware that a few Red-tailed Hawks of other races can look bright white at times, especially when flying over snow cover. Light Harlan's typically have lightly marked undersides with light to moderately marked bellies, and white throats **appearing much like Eastern birds** in this respect. The upperside of all Harlan's adults is dark brown with white mottling limited to the scapulars, and the head lacks the golden nape of other races. **Light Harlan's juveniles often have whitish wing panels and prominent white mottling along the upperwing coverts.** The head often shows white markings around the eyes and white streaking on the crown, which makes the bird appear pale-headed at times. Some are particularly white-headed, especially juveniles. Many Harlan's adults lack banding to the remiges, which can sometimes be seen at a fair distance; this is rare on other races.

Light Harlan's that are lightly marked underneath or show whitish heads and tails may be impossible to tell from Krider's at a distance. **Harlan's adults lack the extensive whitish mottling and rufous tones on the** upperwings that Krider's show. Also, Harlan's adults rarely lack a bellyband. Both, Krider's and light Harlan's juveniles can have whitish wing panels and extensive white mottling along the upperwing coverts. Krider's juveniles can have a whitish tail with dark bands throughout; **Harlan's juveniles rarely if ever have a whitish tail.** Harlan's juveniles can be bright white underneath similar to adults; Krider's often show a buffy wash to the underside.

Dark Harlan's **are blackish or brownish on the underside with varying amounts of white mottling on the chest.** Some Harlan's are solid dark on the body and a few adults have a rufous chest similar to that of Western birds. Sometimes, the carpals on dark Harlan's adults are mottled and appear paler than the underwing coverts at a distance. Dark Harlan's juveniles can exhibit black-and-white tones to their plumage, but are almost identical to dark Western juveniles at a distance, especially birds with streaked undersides. Be aware that dark Western juveniles over snow cover can appear black and white like some Harlan's. **Some dark Harlan's juveniles exhibit white mottling to the upperwings and whitish primary panels appearing speckled on top.** Dark juveniles without this speckled topside are impossible to classify to race at a distance. Although a very uncommon occurrence, adult dark Harlan's are the only buteos to show white on the leading edge of the wings toward the body.

RT 01 - **Light juvenile Red-tailed Hawks** are pale underneath with dark patagials and bellyband (top left). Many **Easterns** (shown) and some Westerns are lightly marked underneath (top right), with **Krider's** almost unmarked underneath with whitish heads (middle left). Many **Westerns** (shown) and some Easterns are heavily marked underneath (middle right). Some juveniles show reddish tails (Eastern, bottom left) or a dark trailing edge to the wings similar to that of adults (bottom right), but show pale wing panels and a less defined dark trailing edge to the wings.

RT 02 - In spring, **juveniles** can have comma-like wing panels due to molt (top left). Head-on, **juveniles** are brownish overall with pale highlights along the front edge of the wings; note the slightly bowed wings and broad chest (top right). Against a white sky (middle left) or when shadowed (middle right), the pale wing panels of **juveniles** may still be obvious. Over snow, **juveniles** of all races can appear bright white underneath similar to Harlan's, and the wing panels can disappear (Western, bottom left). At dusk or dawn, **juveniles** can appear rufous underneath (bottom right).

RT 03 - Typical **Eastern adults** (top left) show pale throats and pale underwing coverts. **Krider's adults** (top right, July) often show whitish heads and tails, and are virtually unmarked underneath. Typical **Western adults** (middle left) are rufous-toned underneath with bold markings to the underside; a few show a whitish chest (middle right). Some **adults** have somewhat dark remiges underneath, appearing slightly two-toned (bottom left); others appear all white in direct sunlight (bottom right); note the reddish tail and dark trailing edge on the wings of **adults**.

RT 04 - Against a white sky (top left) or when shadowed (top right) the rufous tone to the body or tail of **adults** may still be apparent, especially on Western adults (shown); wing molt in fall denotes an adult. Over snow, **adults** of all races appear whiter than usual (Western, middle left). **Adults** at dusk or dawn appear dark rufous below (middle right). **Leucistic** birds have varying amounts of white feathers, but often show belly markings and normal colored flight feathers (bottom left). Head-on, Western (shown) and Eastern **adults** are dark brown with golden napes (bottom right).

RT 05 - Most **dark Western juveniles** are heavily streaked throughout the underside (top left) and may appear identical to solid dark juveniles (top right) in the field. When streaking is less dense on the chest, **dark juveniles** can appear pale-chested (middle left) or pale overall in direct sunlight (middle right). When shadowed, **dark juveniles** can appear dark overall or show a slightly paler chest; note pale wing panels (bottom left). Over snow, **dark Western juveniles** appear black and white (bottom right) and can be impossible to tell from Harlan's.

RT 06 - **Dark Western adults** are dark on the body with a slightly paler chest (top left). When backlit (top right) or shadowed (middle left), the underbody often appears uniformly dark, as some **dark Western adults** are (middle right). Head-on, **dark Western adults** (bottom left) are dark overall, and often lack the golden nape of light adults. Be aware that **light adults** can appear dark underneath when shadowed (bottom right).

RT 07 - **Light Harlan's juveniles** are bright white underneath (top left) and can appear pale-headed at eye level (top right), but are difficult to distinguish at a distance. Most **light Harlan's adults** are bright white underneath with grayish or whitish tails (middle left), but some have reddish on the tail, appearing completely red-tailed overhead (middle right). Head-on, **Harlan's adults** often appear pale-faced (bottom left). Over snow **Harlan's** are particularly white, but may appear similar to other races (bottom right).

RT 08 - Most **dark Harlan's juveniles** (top left) are heavily streaked underneath and often impossible to tell from dark Western juveniles (top right) at a distance. Most **dark Harlan's adults** are dark with white on the chest and a grayish tail (middle left), which appears white when fanned (middle right). **Dark Harlan's adults** can appear reddish-tailed from below; note the white on the chest (bottom left). In direct sunlight, dark birds can appear paler than usual (bottom right, same bird as bottom left).

RT 09 - When backlit, the white chest of **dark Harlan's** can be difficult to see (adult, top left). Over snow, the black-and-white tones of some **Harlan's** are prominent (adult, top right). Some **dark Harlan's juveniles and adults** (shown) are solidly dark on the body (middle left). Head-on, the white chest of **dark Harlan's adults** is often visible (middle right). Some **dark Harlan's adults** (bottom left) have a rufous chest and can be impossible to tell from Westerns. **Leucistic Harlan's** often show normal flight feathers, but can be difficult to identify (dark adult, bottom right).

RT 10 - **Light juvenile Red-tailed Hawks** are brown on top with pale primary panels and varying amounts of upperwing mottling (unknown race, top left). In spring, **juveniles** are paler overall due to fading, especially the head and tail (unknown race, top right). Some **juveniles** show reddish tails (unknown race, middle left). **Krider's and light Harlan's juveniles** can be whitish on the head, tail, and upperwings (unknown race, middle right). **Dark Western juveniles** are darker on top than are light birds (bottom left). **Juveniles** of all races show longer tails and slimmer wings than do adults (Western, bottom right).

RT 11 - **Light Western and Eastern adults** are brown on top with golden heads and rufous tails (top left). In spring, **adults** can be paler overall due to fading (top right). **Dark adults** are darker on top than light adults; note the rufous uppertail coverts that dark adults and some light adults show (middle left). When shadowed, the red tail of **adults** can appear dark (middle right). **Leucistic adults** have varying amounts of white and normal feathers; note the red tail (bottom left). **Dilute plumaged** birds are tan overall (bottom right).

RT 12 - Most **Harlan's juveniles** are brown on top (light, top left) or blackish (dark, top right) with whitish mottling on the upperwings and whitish primaries. **Dark Harlan's juveniles** may lack whitish markings on top (middle left). **Light Harlan's adults** are brown on top with white on the scapulars and a grayish tail (middle right) or a reddish, whitish, or white tail with a rufous distal (bottom left), appearing similar to **Krider's adults** (bottom right, July). Note the whitish head and mottling on the rufous upperwing coverts of Krider's.

RT 13 - **Dark Harlan's adults** are uniformly dark on top and show a grayish tail with rufous distal (top left), a brown or blackish tail (top right), or a whitish tail (middle left). Harlan's with virtually all white tails are similar to some **adult Krider's** (middle right, July), but lack mottling on the upperwings and often show a smudgy, dark tail tip. Some **dark Harlan's adults** show white patches near the neck that other buteos do not show (bottom left). **Leucistic Harlan's** may show Harlan's traits; note the grayish tail and lack of mottling to the upperwings (adult, bottom right).

Ferruginous Hawk
(Buteo regalis)

OVERVIEW

Ferruginous Hawks are large buteos with **long, fairly broad wings, long, broad tails, and stout bodies.** The wings are fairly straight along the trailing edge and taper at the "hands," appearing neatly angular. The wings of Ferruginous Hawks are slimmer at the base than those of Red-tailed Hawks and not quite as smoothly curved along the trailing edge or pointed at the tips as are the wings of Swainson's Hawks. They are deceiving on migration and can look like Swainson's Hawks, Red-tailed Hawks, Northern Harriers, or Turkey Vultures, especially along a ridge. When gliding, the wing tips protrude well past the back of the wings, making them appear "M"-shaped overhead similar to Swainson's Hawks but not quite as long-"handed."

Ferruginous Hawks soar in wide circles with wings held in a pronounced dihedral. When gliding, they hold their wings in a modified dihedral and teeter side to side similar to Northern Harriers but less exaggeratedly. The wing beats of Ferruginous Hawks are **relaxed but stiff and somewhat lofty on the upstroke.** They are slightly quicker on the upbeat than on the downbeat. At times, Ferruginous Hawks display shallow wing beats during powered flight over flat lands.

PLUMAGE

Juvenile light Ferruginous Hawks are brilliant white underneath and essentially unmarked, though they do have faint dark comma-shaped markings at the "wrists" and faint brown spotting to the underwing coverts and leggings. They often **appear completely white underneath at a distance.** Some juvenile Ferruginous Hawks show fairly mottled underwing coverts, but the mottling is much less distinct than on adults. The tail of juveniles appears pale overall from below when folded, but when spread the white base and contrasting dark distal are obvious. The black primary tips on all ages of Ferruginous Hawks are much less prominent than on other buteos. All juveniles are uniform brown (slightly paler and more rusty-brown compared to other juvenile buteos) on top with white primary panels and a white base to the tail. **The white areas on top are noticeable at a distance.**

Adult light Ferruginous Hawks are white underneath similar to juveniles, but **show rufous leggings (which appear as a dark "V" at the belly) and rufous mottling on the underwing coverts.** Some adults have faint rufous leggings that are difficult to discern at a distance. Heavily marked adults can have rufous barring on the belly or throughout the entire underside; the latter is uncommon and makes them appear dark at a distance. The tail of adults appears pale and unmarked from below, but when fanned, typically appears whitish (some birds show a pinkish tail) or **translucent when backlit.** The dark trailing edge to the wings that most adult buteos show is less distinct on Ferruginous Hawks. Some light birds in their first adult plumage are lightly marked underneath and less vibrant in color overall.

From above, adult Ferruginous Hawks are brownish with **vibrant rusty-orange upperwing coverts, obvious white primary panels, and whitish tails with varying**

Ferruginous Hawk

amounts of reddish or grayish mottling. Some individuals have completely gray tails or even red tails (similar to those of adult Red-tailed Hawks). A few adults have faint, whitish wing panels. Adult Ferruginous Hawks appear particularly hooded at eye level or from below due to the contrast between the bright white body and throat and the brown head. However, adults can have pale faces and appear white-headed at a distance as well.

Dark Ferruginous Hawks are brown to rufous-brown underneath; a few are blackish. In good light, *adults* can show a rufous tone to the underside with a slightly darker chest. *Juveniles* are often brownish with a slightly paler chest, but some adults can show this pattern. Although the dark terminal edge to the wings on adult Ferruginous Hawks is less bold than on other buteos, it can help in telling them from juveniles, which show a faint dark terminal edge. Also, juveniles can have a broad, dark tail tip from below that adults do not show. **Regardless, dark Ferruginous Hawks can be extremely difficult to age in flight,** especially when shadowed. **All Ferruginous Hawks have bright white remiges below,** which lack prominent banding, distinguishing them from other dark buteos.

From above, juvenile dark birds are warm brown with pale primary panels. Some dark juveniles lack the white base to the tail that light juveniles show. **Adult dark Ferruginous Hawks are brown to slate-brown on top with faint grayish primary panels.** The primary panels on dark birds are less obvious than those of light birds. Some adult dark Ferruginous Hawks have rufous upperwing coverts, but show considerably darker uppersides overall than light birds. The topside to the tail of dark adults is typically grayish or grayish with whitish or rufous mottling.

FH 01 - Juvenile light **Ferruginous Hawks** are white underneath with faint markings on the underwings and a broad, dark tail tip (top left). Some **juveniles** appear unmarked overall; note the translucent flight feathers when backlit (top right). In direct sunlight, the white underside of **juveniles** is obvious (middle left). When shadowed, **juveniles** can appear dark overall and impossible to age (middle right). Head-on, **light juveniles** appear pale-headed and show a pale leading edge to the wings (bottom left). The pale primaries of **juveniles** are obvious from underneath in certain instances (bottom right).

Rough-legged Hawk
(Buteo lagopus)

OVERVIEW

Rough-legged Hawks are large buteos with long wings and somewhat long tails. In several ways **they resemble a cross between a Red-tailed Hawk and a Northern Harrier.** Their wings and tail are somewhat slimmer than those of Red-tailed Hawks, but broader than those of Harriers. Males are more Harrier-like than females, in that they are slightly smaller with slimmer wings, but telling sexes by shape is very difficult at a distance. Rough-legged Hawks also resemble Ferruginous Hawks, but have a slight bulge along the back edge of the secondaries and lack the angular look and sharply tapered wing tips of Ferruginous Hawks.

Rough-legged Hawks are steady but buoyant in flight and able to gain lift easily. **They hold their wings in a shallow dihedral or modified dihedral** similar to Harriers and are similar in profile at eye level; however, Rough-legged Hawks have bulky chests compared to those of Harriers. From overhead, **Rough-legged Hawks appear "M"-shaped in a glide** similar to Swainson's Hawks, but show broader, less pointed wings. **The wing beats of Rough-legged Hawks are deep but stiff and "wristy."** Rough-legged Hawks possess great stamina; they are able to hover for hours while in search of prey or migrate for long periods using powered flight.

PLUMAGE

Light Rough-legged Hawks are whitish below with black bellies and black carpals. *Juveniles and adult females* are nearly identical in plumage, but adult females show a well-defined dark terminal edge to the wings and dark tail tip from below; the terminal band on the wings and tail on juveniles is ill defined. **Plumage of *adult males* varies from white underneath with dark mottling throughout the entire underbody to white underneath with only a dark bib and narrow carpal patches.** Bibbed males can look like adult Swainson's Hawks, but Rough-legged Hawks always show pale flight feathers. Many adult males show multiple black tail bands, but this is hard to judge at a distance.

Light Rough-legged Hawks are brown on top with pale mottling along the upperwing coverts, and pale, sometimes whitish heads. **Juveniles have pale primary panels** that adult birds lack, but that birds in their first adult plumage can exhibit. **Adult males usually show grayish tones throughout the upperside, whereas adult females may have grayish tones on the back only.** Light Rough-legged Hawks have a **dark tail with a broad, white base,** but some (usually adult females) show white limited to the very base. Many adult males have multiple black tail bands that may be visible from above when the tail is fanned.

There is overlap in plumage between adult females and males. Some adult females have dark bibs, appearing dark overall with a white "necklace"; such heavily marked females lack multiple tail bands and extensive mottling on the underwing coverts. Adult males can look like females or typical males in their first adult plumage or any successive years. Juvenile males and females are identical in plumage. *Dark* Rough-legged Hawks are dark brown underneath with pale flight feathers;

Rough-legged Hawk

some adult males are blackish. Sometimes the "wrists" and bellies are noticeably darker than the rest of the underside, but this is difficult to see at a distance. The flight feathers of adults are more boldly banded than are those of juveniles, giving the underwings a silvery appearance at times. **The trailing edge to the wings on adults shows a broad, dark band, whereas the terminal band on juveniles is less distinct.** The underside of the tail on dark birds appears identical to that of light birds when folded, but **most adult dark birds show multiple tail bands** that are visible from below when the tail is spread.

The topside of dark Rough-legged Hawks is uniformly brownish, lacking pale mottling along the upperwings. Blackish adult males show a dark gray-blue cast to the upperside. The pale wing panels of all juveniles are visible with good views. Juvenile dark birds have pale eyebrows and slightly paler cheeks than do adults, but dark birds in their first adult plumage can also show this head pattern. At a distance, the tail of juveniles is dark overall; adults can show dark tails or have narrow, white bands, but **all dark birds lack the white base to the tail that light birds show.** Some adults (especially males) show a small white patch on the back of the head that juvenile birds lack.

RL 01 - **Juvenile light Rough-legged Hawks** are pale below with a black belly and carpals and a dark, smudgy tail tip (top left). When backlit, the paler primaries of **juveniles** can be obvious (top right). **Adult light females** are similar to juveniles, but show a defined tail tip and dark trailing edge to wings (middle left). When backlit (middle right) or shadowed (bottom left), markings of **adult females** may be obscured. Head-on, **adult light females** typically show pale heads; the black belly and carpals may be visible (bottom right).

RL 02 - At eye level, **juveniles** (top left) and **adult females** (top right) appear similar; note the dark trailing edge to the wings on an adult. All Rough-legged Hawks show pointed wing tips when flapping. **Adult light males** can have pale bellies with dark chests and mottled underwings (middle left), and look particularly pale when backlit (middle right). Some **adult males** show defined bibs similar to those of Swainson's Hawks (bottom left). Over snow, the dark markings on light birds are highlighted; note the dark carpals and bib on this **adult male** (bottom right).

RL 03 - Some **adult males** are similar to adult females, but show less defined carpals, mottled chests and underwings, and multiple tail bands (top left). Some **adult males** are extremely marked on the underwings with darker carpals and a white chest band (top right). Heavily marked **adult males** can appear dark underneath with a paler chest band when shadowed (middle left) or a completely dark underbody in poor light (middle right). Adult females may have pale lower bellies while adult males may look female-like, making some birds impossible to categorize (bottom left, right).

RL 04 - **Dark juveniles** (top left) are similar to **dark adults** (top right), but are slightly paler underneath, lack a defined dark tail tip and trailing edge to wings, and often show pale primary panels. In overcast skies, dark birds appear blackish on the body (juvenile, middle left). Adult males and females can be identical in plumage, but some **adult males** show blackish bodies (middle right). In direct sunlight, the black carpals can contrast the slightly paler underwing coverts on dark birds (adult female, bottom left). Black **adult males** can show silvery primaries in direct sunlight (bottom right).

RL 05 - **Adult dark males and females** can both show multiple tail bands and be impossible to sex (top left), but dark adults lacking multiple tail bands are typically **females** (top right). **Light juveniles** are brown on top with white at the base of the tail and paler primaries (middle left), but in poor light can appear dark (middle right). Head-on, **light juveniles** show a pale chest, head, and a pale leading edge to the wings (bottom left). In spring, **light juveniles** are paler on top than usual due to fading; note the pale wing panels (bottom right).

RL 06 - **Light juveniles** (top left) and **adult light females** (top right) are similar on top, but the pale wing panels and mottling on the upperwings of juveniles are more distinct. **Adult light males** are similar to females, but show a grayish tone on the back (middle left) and may show a grayish head and secondaries (middle right). All Rough-legged Hawks have a white tail base, but this is limited on some birds (adult female, bottom left). Head-on, the pale head of most light Rough-legged Hawks is evident (adult female, bottom right).

RL 07 - **Dark juveniles** are darker on top than are light juveniles and lack a white tail base; note the pale wing panels (top left). In spring, dark birds can appear paler than usual, especially juveniles (shown, top right). **Dark adults** show darker uppersides than do dark juveniles. **Adult females** have a dark tail with a blackish tip (middle left), or multiple tail bands like **adult males** (middle right), but lack a grayish sheen to the back. **Light birds** in poor light (bottom left) can appear dark on top. **Dark birds** in poor light may appear jet black on top (bottom right).

FALCONS

OVERVIEW

With long, narrow, pointed wings, falcons are built for speed. They are able to chase down prey (including other birds) in flight, and can migrate long distances using powered flight. Falcons are extremely steady fliers, excluding Kestrels, which are buoyant, delicate, and not nearly as swift as other falcons. American Kestrels are also the only falcons to hover as they hunt. All falcons hold their wings flat or slightly bowed in all postures. American Kestrels and Merlins are small, only slightly larger than jays, whereas Peregrine and Prairie Falcons are larger than crows. Gyrfalcons range from nearly equal in size to Peregrines to as large as Red-tailed Hawks.

Although size can be difficult to judge in the field, American Kestrels and Merlins display rapid wing beats and quick turns associated with smaller birds. Peregrine Falcons, Prairie Falcons, and Gyrfalcons flap comparatively more slowly and make wider, swinging turns. When gliding high above on migration, **Peregrine and Prairie Falcons** often fly faster than other raptors in the sky. This is especially useful for telling Peregrine and Prairie Falcons from Northern Harriers, which can look particularly falcon-like when gliding. Each species of falcon displays distinctive wing beats. American Kestrels appear to lack power, **flipping their wings back in a quick but soft, sweeping fashion.** The wing beats of Merlins are quick, stiff, and seemingly powerful. On occasion, both species exhibit courtship displays during spring migration, where they flap in a stiff, shallow, rapid fashion with wings severely drooped or cupped (bowed at the "wrists"). Peregrine Falcons display powerful, fluid, whip-like wing beats. Prairie Falcons flap in a similar manner, but their wing beats are somewhat stiffer, shallower, and quicker. The Gyrfalcon's flap is similar to that of the Prairie Falcon's, but more labored.

Of the falcons, Kestrels are the slimmest overall, with narrow wings and tails and relatively small heads. The wings are nearly equal in width throughout their length and taper less sharply at the tips than those of other falcons. Merlins are stockier than Kestrels, especially the chest and base of the wings. Peregrine and Prairie Falcons show extremely long wings, which are broad at the base, and show broad chests and heads. The Gyrfalcon's are the broadest, least sharply pointed wings of the falcons. All falcons show long, pointed wing tips in a glide, exhibiting a shallow "M" shape. Kestrels show a smooth curve to the leading edge of their wings, lacking sharply cornered angles. Excluding Kestrels, female falcons are obviously larger and have longer wings than do males. American Kestrels and Merlins are the only falcons to display obvious sexual plumage dimorphism.

PLUMAGE

American Kestrels are pale underneath with dark streaking on the body and typically appear pale overall; they are orangey on top unlike other falcons. Merlins are heavily streaked underneath and typically appear dark overall, except for Prairie Merlins and adult male Taiga Merlins, which are quite pale underneath and similar to American Kestrels. Merlins are brown on top with black-and-white tail bands; adult males are bluish. Prairie Falcons are whitish underneath with dark streaks on the body and dark wing pits and wing linings, appearing pale overall. They are dark tan overall on top. Peregrine Falcons are variable in plumage. Some juveniles are extremely heavily streaked or marked underneath appearing dark overall, while others are more lightly marked. Adults are white underneath with dark barring on the belly and underwings, and bluish on top with slightly darker flight feathers and head. Gyrfalcons are the most variable in plumage, being nearly completely white to completely blackish overall. All falcons acquire adult plumage after their first molt at about one year old, except Kestrels, which attain adult plumage their first fall.

American Kestrel

(Falco sparverius)

OVERVIEW

American Kestrel is the smallest North American falcon (about the size of a jay), appearing **dainty, slim overall, and lively and nimble in flight**. Their buoyancy in flight is unique among falcons, and they are **fluttery and turbulent at times**, especially on windy days, while other falcons cut through the air in jet-like fashion. They rise quickly while soaring, and their **wing beats are easy and swept back, lacking stiffness and power.** Even when flapping purposefully, American Kestrels move relatively slowly. Kestrels flap and glide intermittently more often than other falcons, and are the only falcons to hover while hunting.

Kestrels have **slim, pointed wings, narrow tails, and slim bodies** compared to those of other falcons. At eye level, **the wings of Kestrels are smoothly curved but hunched timidly at the shoulders.** When Kestrels are soaring high above, their wings can look particularly lengthy, similar to the look of Peregrine Falcons. However, the wings are slim at base, the head appears small, and the tail is slim and does not taper toward the tip. With wings and tail fully fanned, Kestrels sometimes appear unusually stocky and similar to Sharp-shinned Hawks. Although similar in size to Merlins, Kestrels typically appear pale overall underneath, whereas Merlins typically appear dark. Coloration can be helpful when telling Kestrels from Merlins in flight. However, be aware that **Kestrels can appear dark underneath on overcast days or when shadowed.** A common mistake birders make is identifying dark Kestrels as Merlins.

PLUMAGE

American Kestrels are overall pale underneath. *Male* Kestrels have white spots along the trailing edge of the wings that appear translucent when backlit; *females* have buff-colored, less obvious spots. The **underwings of males are checkered black and white, appearing silvery at a distance,** in contrast to the paler body. **Female Kestrels appear uniformly pale underneath at a distance.** Females are streaked on the body; males are spotted and have an orange breast. Juvenile males lack the orange breast and are lightly streaked underneath until late fall. A few juvenile males are fairly heavily streaked, but never appear as prominently marked as Merlins.

Kestrels are dark orange on top, with **males having blue upperwings, black remiges,** and a broad black tail tip (which can be obvious from underneath); females have a narrow black tail tip and multiple thinner black bands throughout the tail that are difficult to see at a distance. The outer tail feathers of male Kestrels may have multiple bands, and when folded, the tail can look completely banded underneath similar to that of a Merlin. American Kestrels often look pale-headed at a distance, distinguishing them from Merlins (except Prairie Merlins). Some adult Kestrels show flight feather molt during fall, but juveniles only show flight feather molt in spring.

American Kestrel

AK 01 - **Female American Kestrels** appear uniformly pale underneath in good light (top left) or when backlit (top right). **Female** Kestrels appear whiter than usual in direct sunlight (middle left) or over snow (middle right). When shadowed, Kestrels may appear uniformly dark and impossible to sex (bottom left). Blunt wing tips due to primary molt (denotes adult) is the only way to age females in fall (bottom right). Note the long, narrow wings and tail characteristic of all Kestrels.

AK 02 - **Male** Kestrels are pale below with darker underwings. **Juvenile males** in fall have dark streaking on the chest (top left); **adult males** (top right) have an orange chest; note the broad black tail tip that all males show. In good light, the underwings of **males** appear silvery and darker than the body (adult, middle left). Some **juvenile males** are heavily streaked on the body (middle right). When shadowed, males may be difficult to sex; white spots on the back edge of the wings are more prominent on males than on females (bottom left). In direct sunlight, males may appear completely whitish (bottom right).

AK 03 - **Female** Kestrels are orange on top with blackish remiges (top left); **males** have blue upperwing coverts and a broad, black tail tip (top right). In poor light, Kestrels may appear dark overall (middle left). All Kestrels flip their wings back in a sweeping motion when flapping (middle right). Head-on, **adult males** show an orange chest (bottom left); **females** and juvenile males are buff throughout (female, bottom right).

Merlin

Merlin

(Falco columbarius)

OVERVIEW

Merlins are small birds, only slightly larger than Kestrels, but **stockier overall and fly with power, speed, and stability.** They are **level fliers with quick, forceful wing beats** able to propel them to high speeds in seconds, often giving birders only an instant to identify them. Compared with Kestrels, Merlins have stout chests, broader-based wings, and slightly broader, shorter tails. Compared to the larger falcons, Merlins have shorter wings and slim, short tails that are typically square-tipped when folded, lacking a taper toward the tip. Like all falcons, Merlins hold their wings flat or slightly drooped at all times.

Merlins are feisty and often harass birds much larger than them. They frequently hunt during migration, and while doing so males seemingly attempt to mimic songbirds by flipping their wings back in a flashing, rapid, intermittent fashion. Some female Merlins attempt to deceive prey by exhibiting slow, lofty wing beats before surprising them with a quick burst of speed. Females are slightly larger than males and have proportionately longer wings, but telling sexes in flight based on shape can be very difficult. Merlins are known to migrate at sundown and beyond, when it is too dark to see age and sex characteristics.

There are three races of the Merlin— Taiga (*F.c. columbarius*), Prairie (*F.c. richardsonii*), and Black (*F.c. suckleyi*)—and each race is variable in plumage. The Taiga race occurs throughout most of North America, but is the only race seen in the Eastern half. Prairie Merlins occur primarily in the Western half. Black Merlins occur along the West Coast, but may be found in the Intermountain West on occasion.

PLUMAGE

Taiga Merlins are heavily streaked underneath with a paler throat. ***Adult females and juveniles*** have dark brown, dense streaking and often appear dark in the field. ***Adult males*** are streaked with rufous or rufous-brown with a more obvious whitish throat, appearing more Kestrel-like underneath. The "wrists" and leggings of most adult males are yellowish, but this is noticeable only in adequate light.

Adult female and juvenile Taiga Merlins are dark brown on top. Adult females are typically more slate-brown than juveniles, but **the two are often impossible to tell apart in flight.** Juvenile male Taiga Merlins may be brownish or dark bluish-gray on top. Adult male Taiga Merlins are bluish on top with blackish primaries. **Merlins usually have distinct pale bands on the tail and a prominent white tip**, especially adult males. The tails of some Taiga Merlins lack bands but have a dark sub-terminal band. Some Taiga Merlins from the Great Lakes region are nearly as dark as Black Merlins. Intergrades between Taiga and Prairie Merlins and between Taiga and Black Merlins occur, which also makes distinguishing the race of some impossible.

Prairie Merlins are considerably paler overall than Taiga Merlins, and similar in plumage to American Kestrels, making identification difficult at times. *Juveniles and adult females* show whitish undersides with pale rufous-brown streaking and often appear pale-headed from below. They are rufous-brown on the upperside, but are more uniform in tone, similar to that of Prairie Falcon, whereas Kestrels show a contrast between the pale upperwings and darker primaries. *Adult male* Prairie Merlins are extremely pale below with faint rufous barring on the body; some are heavily streaked and appear more similar to Taiga birds. The topside of adult male Prairie Merlins is pale blue with blackish primaries. All Prairie Merlins (especially adult males) have well defined tail bands with a prominent broad, dark sub-terminal band.

Black Merlins are dark overall compared with other races of Merlin. *Juvenile and adult female* Black Merlins are blackish on top and extremely heavily marked underneath, appearing solid dark in the field. Black Merlins typically lack the "moustache" markings and whitish throats shown by other races. *Adult male* Black Merlins are slightly less marked underneath than juveniles and adult females and can show a buffy tone, but sexing Black Merlins from the underside is very difficult in flight. The upperside of adult male Black Merlins is dark blue with black primaries. Black Merlins of all ages and sexes often show dark tails that lack well-defined bands.

ML 01 - Juvenile and **adult female Taiga Merlins** are pale underneath with significant dark streaks on the body, dark underwings, a pale throat, and a banded tail (top left). When shadowed, **Taiga** Merlins appear uniformly dark (top right), but slightly paler when backlit (middle left). **Juvenile and adult female Prairie** Merlins are pale underneath with rufous-brown streaking (middle right, bottom left). **Juvenile and adult female Black** Merlins appear nearly solid blackish underneath (bottom right).

ML 02 - Adult male Taiga Merlins are slightly paler underneath than are females and have golden leggings and "wrists" (top left). **Adult male Prairie** Merlins are pale underneath with rufous body markings; note the golden leggings (top right). Merlins often migrate at sundown and are difficult to age or sex in low light (Taiga, middle). Head-on, Merlins show dark heads with pale throats (Taiga, bottom); note the relatively stocky wings and chest.

ML 03 - Juvenile Taiga Merlins are brownish above with multiple tail bands (top left); some have only one dark band at the tip (top right). **Juvenile male Taigas** can be dark slate blue on top (middle left), similar to the slate color of **adult females** (middle right). **Juvenile and adult female Prairie** Merlins are orangey-brown above, similar to female Kestrels, but have banded tails and do not show blackish primaries (bottom left). In poor light, **Prairie** Merlins may appear dark on top (bottom right).

ML 04 - **Adult male Taiga** Merlins are dark blue on top with blackish remiges and well-defined tail bands (top). **Adult male Prairie** Merlins are pale blue on top with blackish remiges, well-defined tail bands, and a broad, black sub-terminal tail band (middle). **Adult male Black** Merlins are dark blue (or blackish like juveniles and adult females) on top with black flight feathers and a blackish head; they often lack tail bands (bottom). Note the stocky wings characteristic of all Merlins compared to those of other falcons.

Peregrine Falcon

(Falco peregrinus)

OVERVIEW

Peregrine Falcons are large falcons with long, narrow, pointed wings and powerful chests. **Compared to those of Kestrels and Merlins, the wings of Peregrines are longer and relatively broader at the base, the chest and head are stout, and the tail is long and broad and tapers toward the tip** (when folded). Peregrines display powerful, fluid, whip-like wing beats with which they can accelerate to high speeds in seconds. The wings of Peregrines are smoothly curved with a slight bend at the "wrists," making the overall silhouette resemble that of **a "cocked bow and arrow"** when soaring. Peregrine and Prairie Falcons are very similar in shape, but Prairie Falcons show slightly slimmer wings and tails, appearing more Kestrel-like than Peregrines. With adequate views, Peregrines show dark profiles with contrasting pale chests, whereas Prairie Falcons exhibit a pale profile overall with dark wing pits and wing linings.

There are three races of the Peregrine Falcon in North America: Tundra (*F.p. tundrius*), Anatum (*F.p. anatum*), and Peale's (*F.p. pealei*). Tundra birds occur across North America and are the most common of the races. Anatum birds are seen primarily in the West. Peale's race originates from the islands off Alaska and British Columbia, and is seen along the West Coast in fall and winter. Some Peale's birds move great distances east and can be seen along the East Coast at times. Female Peregrines are larger and have broader wings than do males, but telling sexes of Peregrines at a distance is difficult.

PLUMAGE

Juvenile Peregrines are buff below with dark streaking on the body, checkered underwings, and a pale throat, appearing dark underneath when shadowed but pale underneath in direct sunlight or over snow cover. By spring, the buff color on the underside fades to whitish. From above, juveniles are slate-brown, sometimes showing slightly darker primaries and an indistinct tail with a white tip.

Adult Peregrines are pale below with faint dark barring, a white chest, a blackish hood, and checkered underwings. **Adults are bluish on top with a blackish head** and blackish primaries, sometimes appearing two-toned overall. Some darker adults can be blackish-blue overall on top, while others have a pale blue "rump" that contrasts with the upperwings and tail. Adults often appear pale-chested, while juveniles are pale-throated at a distance.

Tundra juveniles are the least heavily marked underneath of the three races and the most likely to show pale foreheads. *Anatum* birds are typically more heavily marked than Tundra birds, and often show a strong rufous wash to the underside and dark foreheads. *Peale's* race is extremely heavily marked underneath and dark above similar to dark Gyrfalcons. All Peregrines have dark "sideburns" on the head. The pale forehead of Tundra juveniles and some Anatum Peregrines is often visible at a distance. Adult Tundra and Anatum birds typically have unmarked chests and throats, whereas Peale's birds may have barring throughout the underbody. There is plum-

Peregrine Falcon

age variation in all races and telling them apart in flight, especially Tundra and Anatum birds, can be impossible. Some Peregrines in the East resemble Peale's individuals, possibly the result of mixed origin birds used during reintroduction efforts.

PG 01 - All **juvenile Peregrines** are moderately (top left) to heavily (top right) streaked underneath, with most showing pale throats. Many **Tundra juveniles** have pale crowns and cheeks (middle left). Over snow, **juvenile** Peregrines can look overall unusually pale (middle right). When shadowed, **juveniles** can appear uniformly dark underneath (bottom left). Only adults show flight feather molt in fall, but **juveniles** (shown) and some adults may show molt in spring (bottom right).

PG 02 - Adult Peregrines are heavily barred below with paler chests and throats and a black head (top left, right). Over snow, **adults** appear pale grayish underneath with black heads and dark remiges (middle left), as opposed to uniformly dark when shadowed (middle right). Head-on, **adults** are dark-headed and pale-chested (bottom left); **juveniles** are paler-headed with whitish throats (bottom right).

PG 03 - Juvenile Peregrines of all races are brown (top left) to slate-brown (top right) above with many having pale crowns. **Adult** Peregrines are dark blue on top with black remiges and a black head (middle left, right); note the brown juvenile inner primaries and tail feathers on many birds in their **first adult plumage** (middle right). Some **adults** are relatively darker overall and more uniform in tone above (bottom left). In poor light, **adults** can look black on top (bottom right).

Prairie Falcon

Prairie Falcon
(Falco mexicanus)

OVERVIEW

Prairie Falcons occur primarily in the West. They are large falcons, about the size of Peregrines, and nearly as impressive in speed. They also are similar in shape and flight style, except Prairie Falcons have somewhat slimmer wings and tails, at times **resembling a "big Kestrel."** While a Prairie Falcon's shape can appear somewhat Kestrel-like, the wings, tail, chest, and head are relatively broader.

The flight style of a Prairie Falcon is also like that of a Peregrine: steady, strong, and fast when flapping with intent. Prairie Falcons flap in a manner similar to Peregrines; however, **their wing beats are slightly stiffer, shallower, and almost Merlin-like** but much less swift. Female Prairie Falcons are larger and have slightly broader wings with blunt wing tips, whereas the wing tips of males are often sharply pointed; however, telling sexes of Prairie Falcons at a distance is difficult.

PLUMAGE

Prairie Falcons are *whitish underneath with dark wing pits and wing linings*. Even in poor light, the dark wing pits of Prairie Falcons are usually obvious. *Adults* are spotted on the underbody, whereas *juveniles* typically show dark streaks (some juvenile males may be spotted). Adult males and females are often identical in plumage, but some males are lightly marked on the body; most females are more heavily marked, appearing slightly darker at a distance. **It is often impossible to age or sex Prairie Falcons at a distance**, especially without considerable experience. **The pale flight feathers of Prairie Falcons can appear translucent**, especially against a blue sky, giving the wings a slimmer appearance than normal and making the tail appear pinkish.

Prairie Falcons are brown above; they are paler than Peregrine Falcons but slightly darker than Prairie Merlins. Juveniles are slightly darker overall than adults. The tail of adults is often paler than the upperwings, showing a slight contrast on top.

PR 01 - All **Prairie Falcons** are whitish underneath with dark wing pits and underwing linings (top left). **Juveniles** are typically streaked underneath (top right); **adults** are spotted and often appear somewhat paler on the body (middle left), but some **adults** (often females) are heavily marked; note the primary molt in fall (middle right). Over snow, Prairie Falcons are whitish underneath (juvenile, bottom left), but when shadowed, may appear dark overall (bottom right). Telling ages in flight is often difficult.

PR 02 - When backlit, **Prairie Falcons** can show pinkish tails and translucent remiges (top left). At sundown, Prairie Falcons can look unusually buffy underneath (top right) or appear to lack field marks (middle left). Head-on, Prairie Falcons are pale-headed with whitish chests (middle right). **Juveniles** are uniformly warm brown above (bottom left); **adults** are slightly paler on top, especially the tail (bottom right).

Gyrfalcon

(Falco rusticolus)

OVERVIEW

Gyrfalcons are the largest, most rare, and the most sought after North American falcon. They are often described as 'big, dark falcons," but actually are highly variable in plumage, ranging from almost completely white to completely blackish. In flight, male Gyrfalcons, which are considerably smaller and slimmer-winged than are females, often do not look larger than Peregrine Falcons. **The wings and, to a lesser extent the tails, of Gyrfalcons are slightly broader than those of Peregrine and Prairie Falcons,** and the wing tips are more rounded; this is most apparent on Gyrfalcons when they are soaring or in a shallow glide. Gyrfalcons have bulky chests and backs, resembling a **"football with wings."** Peregrines and Prairies are slimmer. **The wing beats of Gyrfalcons are heavier, shallower, and less whip-like than those of Peregrines, and more like those of Prairie Falcons.**

At times, escaped falconry birds are seen in the wild. Some of these birds are hybrids involving Gyrfalcons. Birds commonly bred with Gyrfalcons are Peregrine, Saker, Lanner, and Prairie Falcons. These hybrids can look like a mix of two species or nearly identical to one species, making some impossible to tell from pure Gyrfalcons. Falconry birds will often have jesses or rings on their legs, but these can be difficult to view in the field.

PLUMAGE

Gray Gyrfalcon is the most common form in Gyrfalcon and the most similar in appearance to Peregrine Falcons with which they are often confused. *Juvenile gray* Gyrfalcons and juvenile Peregrine Falcons both show dark streaking underneath and dark brown uppersides. **Although Gyrfalcons are thought to be darker underneath than are Peregrines, many juvenile gray birds are less heavily marked** and less buffy than most Peregrines. Juvenile gray Gyrfalcons are slightly paler brown on top as well; however, **they are difficult to tell from juvenile Peregrines at a distance by plumage alone.** Rarely, juvenile gray Gyrfalcons have dark mottling along the underwing linings similar to those of Prairie Falcons, but these markings are less bold than on Prairie Falcons, and the dark wing pits of Prairie Falcon are absent on Gyrfalcons. *Adult gray* Gyrfalcons are similar in plumage to adult Peregrines. They are paler overall and evenly spotted underneath lacking the white-chested look of adult Peregrines, and paler gray on top. The dark grayish-brown head lacks the distinct blackish hooded look of Peregrines.

White Gyrfalcons are white overall with blackish mottling along the upperside and blackish spotting below. *Juveniles* are similar to *adults,* but are more heavily marked on top and more heavily streaked below. Some adults (especially males) are almost pure white. All white Gyrfalcons have dark tips on the outer primaries, a trait that only appears to show on Prairie Falcons when their wings are drawn in during a steep glide.

Dark Gyrfalcons generally appear solid dark overall with no distinct field marks. *Juveniles* are extremely heavily streaked underneath or sometimes nearly solid dark; *adults* have faint pale barring to the underside. There is overlap in plumage between gray and dark Gyrfalcons; some birds appear darker than the typical gray bird while others are paler than the normal dark bird.

GY 01 - Juvenile gray Gyrfalcons are heavily streaked underneath with pale cheeks (top left). Juvenile and adult (shown) **dark** birds are nearly solid dark with paler flight feathers (top right). **Adult gray** Gyrfalcons are pale underneath with dark spots on the body and a dark crown (middle left). Some **gray adults** are nearly un-marked underneath (middle right), appearing similar to **white adults** (bottom left). **White juveniles** are white above with black barring on the upperwings (bottom right), and white below with black streaks on the body.

GY 02 - Juvenile gray Gyrfalcons are brown on top with slightly darker remiges (top). **Gray adults** are grayish above with slightly darker remiges (middle) and appear similar to Peregrine Falcons, but lack the black head. At eye level, **adult Goshawks** can show pointed wings like falcons; note the black head with white eyeline (bottom). All Gyrfalcons show broad, less pointed wings and broader chests compared to those of other falcons.

VULTURES, OSPREY, EAGLES

Black Vulture
(Coragyps atratus)

OVERVIEW

Black Vultures are large, black birds, appearing only slightly smaller in flight than Turkey Vultures. **They are stocky overall with broad, squared-off wings, small heads, and very short, square-tipped tails.** This is the only raptor in North America whose feet sometimes project beyond the tail. The Black Vulture's flat-winged profile appears similar in shape to that of Bald Eagle when soaring at eye level, but Black Vultures are overall stockier in comparison with smaller heads and shorter wings and tails. Black Vultures soar on flat wings that arch forward (Turkey Vultures and Bald Eagles hold their wings straight across), or with a slight dihedral. They glide with a modified dihedral, as opposed to Turkey Vultures, which typically show exaggerated dihedrals. **The wing beats of Black Vultures appear anxious, quick, and shallow,** and are very different from those of eagles and Turkey Vultures. Black Vultures often flap in short bouts interspersed with glides, giving a distinct accipiter impression. Black Vultures appear steady in flight most of the time, but can appear slightly unstable during strong winds. Black Vultures often soar in groups with Turkey Vultures.

PLUMAGE

Black Vultures are black overall with pale, almost silvery outer primaries. Against overcast skies, the outer primaries can appear dark and the wings lack any contrast. From above, the pale outer primaries almost always contrast with the rest of the wings, even in poor light. *Adults and juveniles* are similar in plumage; sexes are identical. Black Vultures showing wing molt in autumn are adults.

Black Vulture

BV 01 - Juvenile and adult Black Vultures are black overall with pale, silvery outer primaries that can be less distinct when shadowed (top left) or in poor light (top right), but obvious in direct sunlight (middle left) or over snow cover (middle right). From above, the pale outer primaries of Black Vultures are visible in almost all conditions (bottom left). Only **adult Black Vultures** show flight feather molt in fall (bottom right). All Black Vultures have relatively stocky, squared-off wings, short, squared-off tails, and small heads.

Turkey Vulture

Turkey Vulture
(Cathartes aura)

OVERVIEW

Turkey Vultures are easily identified, even at considerable distances. **They are large birds (slightly smaller than eagles) with long, broad wings that lack a noticeable taper** and long tails that are wedge-shaped or rounded. **The heads of Turkey Vultures are small compared with those of other raptors.** Despite their size, Turkey Vultures are buoyant and graceful in flight, and **fly in a wobbly manner, constantly swaying from side to side,** making them distinctive among raptors. Turkey Vultures appear to move slowly in flight and soar lazily in wide circles. **They hold their wings in a strong dihedral or modified dihedral.** Turkey Vultures exhibit lofty, easy wing beats that end abruptly on an upstroke. Turkey Vultures are social and can often be seen in large kettles on migration, especially in the East and in Texas.

PLUMAGE

Turkey Vultures are blackish underneath with paler, or silvery flight feathers, giving them a two-toned appearance. In poor light, the flight feathers of Turkey Vultures look uniformly dark. **They appear completely blackish on top;** some adults show minor fading along the upperwings (unlike the obvious fading that eagles can exhibit), but still appear dark overall at a distance. The head (which lacks feathers) and bill are dark on *juveniles*; *adults* have a reddish head and white bill at about two years old. Birds from one to two years old (*sub-adults*) show an intermediate coloration on the head and bill between that of juveniles and adults. Head color is difficult to see at a distance, but the whitish bill of adults is sometimes obvious on overhead birds. Males and females are identical in appearance.

TV 01 - **Turkey Vultures** are black with paler flight feathers, appearing two-toned underneath when shad-owed (top left) or backlit (top right), but strikingly two-toned over snow. Note the white bill of an **adult** (middle left). In poor light (middle right) or when seen head-on (bottom left) vultures appear completely black. Only non-juveniles show flight feather molt in fall (bottom right).

TV 02 - Turkey Vultures are black above, with **adults** showing red heads (top left). At times, Turkey Vultures briefly flex their wings in flight (top right). Turkey Vultures (middle left) are similar to adult Golden Eagles (middle right), but appear smaller-headed and have less tapered wings. Turkey Vultures missing their tails (bottom left) can be confused with Black Vultures (bottom right), but show longer wings and pale remiges throughout the wings.

Osprey

Osprey

(Pandion haliaetus)

OVERVIEW

Ospreys (also known as "Fish Hawks") are distinctive in shape and plumage, which can make them easy to identify, even at considerable distances. They have **extremely long, narrow wings that are held slightly drooped in all postures, and form a distinct "M" shape in a glide.** Ospreys have narrow chests compared to those of vultures and eagles. They are extremely steady fliers, soaring in wide circles and gaining lift easily. They readily use powered flight, and can do so for long distances. **The wing beats of Ospreys are stiff, somewhat shallow, labored, and seemingly mechanical.** When flapping into strong winds, Ospreys appear to struggle as their heads and chests bob up and down.

PLUMAGE

Ospreys are brilliant white underneath with blackish flight feathers and "wrists." Even in poor light, the bold black-and-white underside is obvious from below. *Males and females* are similar in plumage, but adult males tend to lack the faint, streaked bib on the chest that most adult females and juveniles exhibit. The streaking on the chest can be difficult to judge at a distance. Most *juveniles* show a rufous wash to the underwings, but some adults show this as well. Because most juvenile Ospreys do not return to North America in their first spring, **migrating Ospreys in spring are typically adults.**

The upperside of Ospreys appears black with a white head and dark eye-line. Juveniles show pale edges to the upperwing coverts, creating a mottled appearance, but this mottling is difficult to see at a distance. Juveniles are likely to show streaking on the crown, which makes the head appear less brilliant than that of adults. Ospreys have blackish tails with faint white bands, but from below their tails look pale or even pinkish when fanned and backlit.

OS 01 - **Ospreys** are brilliant white below with dark flight feathers (top left). In overcast skies (top right), they can appear overall darker; note the "M" shape in a glide. When backlit, the tail can appear whitish or pinkish when spread (middle left). When seen head-on, Ospreys exhibit a white head and long, bowed wings (middle right). From above, Ospreys appear similar to adult Bald Eagles, but Ospreys lack a white tail. **Juvenile** Ospreys show pale mottling on the upperwings (bottom left), while **adults** do not (bottom right).

Bald Eagle

(Haliaeetus leucocephalus)

OVERVIEW

Bald Eagles are **large, blackish birds with extremely long, somewhat broad wings, and large, protruding heads.** The wings of Bald Eagles taper (or pinch in) less toward the body than do those of Golden Eagles. The wings of Bald Eagles up to about two years old are slightly broader than those of older birds, although this is difficult to judge without practice. Bald Eagles appear **slow moving and steady in flight, soaring in wide circles, and exhibit stiff, lofty, labored wing beats.** Bald Eagles typically **hold their wings flat or slightly drooped** in all postures. Bald Eagles show several plumages before reaching adulthood at about five years old, and classifying immature birds to a specific age is challenging. Females and males are identical in plumage.

PLUMAGE

Juvenile Bald Eagles are **blackish underneath with white wing pits and varying amounts of white on the underwings.** Be careful: from a side angle the white wing pits look like the white belly of sub-adults. The tail is indistinct when folded, but appears pale below with a darker tip when spread. **By winter, the belly fades to pale brown** or near white on some birds. The two inner primaries of juveniles have pale tips that appear as small, translucent "windows" in the wings (sometimes appearing as missing feathers or gaps in the wings). Sub-adults may show small "windows," but their primaries have dark tips that fill the gaps shown by juveniles.

The topside of juvenile Bald Eagles is blackish with little to fair amounts of white mottling on the tail. By fall, the **back and upperwings fade to brown, and contrast with the blackish remiges, appearing two-toned** (similar to Swainson's Hawk)—a trait vultures and Golden Eagles do not show. By spring, the back and upperwing coverts of juveniles can fade to almost whitish.

Sub-adult Bald Eagles one to three years old are similar in plumage to juveniles, but most **have white bellies, white on the back, paler heads, and uniform blackish upperwings.** The white bellies of sub-adults are often obvious, even in poor light. Sub-adult I birds (one to two years old) typically have bright, white bellies (very few are still brown), and retain several juvenile flight feathers, which are longer, **producing an uneven edge to the back of the wings. Sub-adult II and sub-adult III birds (two to four years old) have a smooth edge to the back of the wings,** but may show one or two longer juvenile feathers. The head and tail become whiter each year, but still appear similar to those of juveniles; the back and belly become darker. Most sub-adult II birds (two to three years old) have limited white on the axillaries and underwing coverts, and have either white or blackish bellies. Compared to sub-adult I birds, the backs of sub-adult II birds typically have less white mottling, and the head is often whiter. Because plumages of one- to three-year-old birds can overlap (partly because some eagles molt at different rates than others), these birds are often impossible to age more specifically at a distance.

Sub-adult III, or transition, Bald Eagles (three to four years old) have almost completely dark bodies with limited white

Bald Eagle

blotches on the underwings and wing pits, but show significant variation in plumage among individuals. The head is white with a smudgy, dark eye-line similar to that of an Osprey, but some can have fairly dark heads. The tail of transition birds is typically white with a dark tip. Transition birds have uniformly dark uppersides, but some have a few white spots on the back. Transition Bald Eagles with mostly dark heads, and a white tail with a dark tip, resemble immature Golden Eagles.

Adult **Bald Eagles are blackish with a brilliant white head and tail.** Although distinct at most times, adults (as with any age) can look uniformly dark in poor light. Against a white background like snow or bright clouds, adults may look headless. Birds in their first adult plumage (four to five years old), sub-adult IV birds, may still have dark streaking in the head, dark spots on the tail tip, and some white in the wing pits or underwings. Some birds in their first adult plumage appear identical to older adults, whereas some birds over five years of age still have dark streaking in the head or dark to parts of the tail tip. Adult Bald Eagles can have a dark, smudgy tail tip due to mud stains that resembles the dark tail tip of sub-adult birds.

BE 01 - Juvenile Bald Eagles are dark below with white wing pits and white along the underwing coverts (top left). Some juveniles have limited white underneath (top right) or extensive white markings (middle left). In spring, juveniles can appear white-bellied due to fading (middle right). When backlit, the tail of juveniles may appear white with a dark tip when fanned (bottom left). When shadowed, juveniles can appear uniformly dark underneath (bottom right).

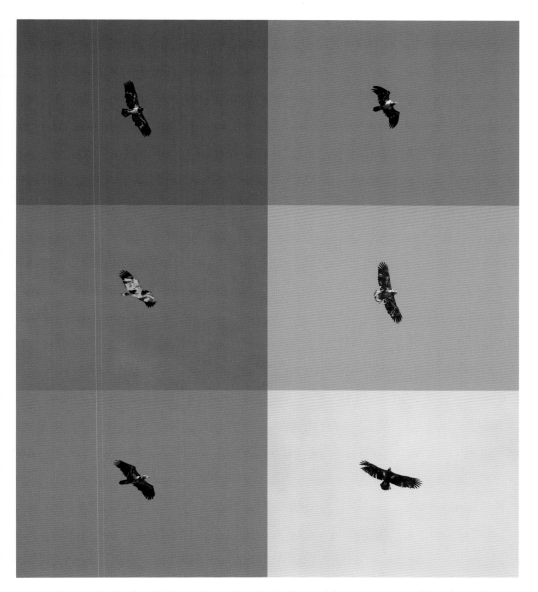

BE 02 - Almost all **sub-adult I** birds are dark with white bellies, and show an uneven trailing edge to the wings (top left) that can be difficult to see in a glide (top right). Some juveniles and **sub-adult I** birds (shown, middle left) have extensive white throughout the underwings. **Sub-adult II** birds (middle right) may have white bellies, but often have somewhat darker bellies than sub-adult I birds and show an even trailing edge to the wings. A few **sub-adult II** birds retain juvenile secondaries that jut beyond the back of the wings or are pale from fading (bottom left, right).

BE 03 - **Sub-adult III** birds are dark underneath with a somewhat whitish head and whitish tail with a dark tip, but show varying amounts of white on the wing pits (top left), underwings (top right), and belly (middle left). When shadowed, the whitish head and tail can be difficult or impossible to discern (middle right). **Juveniles** (bottom left) and **sub-adult III** birds (bottom right) can appear similar in plumage; note the dark head, the brown instead of black belly, the white underwing coverts, and the pale "window" in the wing of the juvenile.

BE 04 - Adult Bald Eagles are black with a pure white head and tail (top left). Birds in their **first adult plumage** (sub-adult IV) often have some white on the underwings or wing pits, or dark on the head (top right). Over snow, the white head and tail of **adults** are gleaming white; note the pale "wrist" commas (middle left). When shadowed, the white head and tail of **adults** may appear dark (middle right) or still appear evident (bottom left). Against a white background or sky, **adults** can appear headless (bottom right).

BE 05 - Juveniles are black on top with brown upperwing coverts, appearing two-toned (top left). In spring, the upperwings and back of **juveniles** can fade to near white (top right). **Sub-adult I** birds are uniformly blackish on top with white backs, and show an uneven trailing edge to the wings (middle). A few **sub-adult I** birds lack the white back, appearing juvenile-like (bottom left). **Sub-adult II** birds often have white on the back, but show a pale crown and straight trailing edge to the wings (bottom right).

BE 06 - Sub-adult III birds are blackish on top with mostly white heads and a mix of white and black on the tail (top). Bald Eagles in their **first adult plumage** show a nearly white head and tail with a dark tail tip (middle left), but may show a near completely white tail (middle right); note the white spots on the secondaries. **Adult** Bald Eagles are completely blackish with fully white heads and tails (bottom left). Be aware that mud stains on the tail tip (bottom right) may appear as a dark tail tip.

BE 07 - **Head-on, juveniles** appear completely dark (top left); the white bellies of **sub-adult I** birds may be visible (top right); the head appears dingy on most **sub-adult II** birds (middle left); **sub-adult III** birds typically have mostly white heads, but some appear dusky and much like a Golden Eagle's (middle right); birds in their **first adult plumage** often have white heads with faint dark streaks (bottom left); **adults** show fully white heads (bottom right).

Golden Eagle

(Aquila chrysaetos)

OVERVIEW

Golden Eagles are **large, dark birds with extremely long, somewhat broad wings, which are held in a dihedral when soaring and a modified dihedral or slightly drooped when gliding.** They appear **slow moving and steady** in flight, soaring in wide, lazy circles. Golden Eagles are similar in shape to Bald Eagles, but have smaller heads and the back edge of their wings taper more at the body. Golden Eagles show similarities to buteos and vultures, but exhibit larger heads and longer wings. Golden Eagles display **slow, easy wing beats that end abruptly on an upstroke** into a dihedral. The wing beats of Bald Eagles are stiff, with deep upstrokes that end on a level posture. During light winds along a ridge, Golden Eagles can exhibit shallow, "wristy," intermittent flaps that also end abruptly on an upstroke. The largest females have more labored wing beats and more pronounced dihedrals (sometimes as steep as those of a Turkey Vulture) than males, but it is often impossible to tell the sex of Golden Eagles at a distance.

PLUMAGE

Golden Eagles are dark overall with a golden nape, with immature birds having varying amounts of white in the wings and tail that gradually become darker as they reach adulthood at four to five years of age. It may be possible to see white patches or molt in the flight feathers to categorize eagles as "immature," "non-juvenile," or "non-adult," but it is often impossible to age Golden Eagles to a specific year. This is because **the white (or lack of) on the tail is often concealed in the field, especially from underneath when the tail is folded.** At times, the golden undertail coverts that all Golden Eagles have can be confused for a white tail base. It is often necessary to see the topside of a Golden Eagle to age it accurately. Golden Eagles never show a striking two-toned upperside or white on the back, like immature Bald Eagles do. Instead, they are dark brown overall, with sub-adults and adults showing pale mottling along the upperwing coverts. Be aware that on bright days the pale head of a Golden Eagle may appear similar to the white head of a Bald Eagle, especially in spring, when the head plumage is faded.

Juveniles are dark overall with a **white base to the tail and white patches in the wings.** The white in the tail is typically extensive, covering more than half the tail, but can be restricted to just the base on some birds. The white in the wings is often solid, but may be divided by dark feathers similar to the wings of sub-adults that have already experienced molt. Be aware that some juveniles have little or no white in the wings, so **the amount of white in the wings is sometimes irrelevant in ageing Golden Eagles;** some birds with prominent white on the wings can be older than birds with no white on the wings. Juveniles with extensive white wing patches to the underside of the wings often show small white patches on the upperside of the primaries. The upperwings of juveniles are uniformly brown, lacking the pale mottled upperwing bar that all other ages show. The upperwing coverts and golden nape of all eagles typically fade by spring and appear paler than usual. Juve-

Golden Eagle

niles lack any signs of molt until spring, so any eagle that is actively molting or shows signs of previous molt during fall and winter is not a juvenile.

Ageing sub-adult Golden Eagles is one of the most difficult aspects of raptor identification. Sub-adult I and sub-adult II birds have juvenile-like (white-based) tails, sub-adult III and sub-adult IV birds have adult-like (dark) tails with white patches on each side, although they vary slightly. Some juvenile Golden Eagles replace their remiges with adult feathers during their first molt while others replace them with sub-adult white-based feathers (birds with white as juveniles), so the amount of white in the wings varies greatly within every age and is often irrelevant in ageing sub-adults. On the contrary, sub-adult tail feathers always have a white base, as it takes several years for eagles to acquire a full adult tail (since they molt more slowly than do other raptors).

Sub-adult I (one to two years old) Golden Eagles tend to retain most of their juvenile feathers, and **appear nearly identical to juveniles**; the trailing edge of the wings does not appear irregular as it does with sub-adult I Bald Eagles. Some upperwing coverts on sub-adult I Golden Eagles are replaced, with the new pale feathers forming a narrow, mottled upperwing bar. However, the upperwing bar on sub-adult I birds is sometimes very limited. The upperwing coverts of juveniles fade by spring, forming a broad, pale area, not a narrow wing bar. Although assessing molt can be difficult in the field, it can prove useful in ageing sub-adult Golden Eagles with significant white in the wings or tail. Sub-adult I birds actively molting in fall show molt along the inner primaries, which appears as a gap along the back edge of the wings. The central tail feathers may be actively molting as well.

Sub-adult I birds not actively molting are difficult to age.

Sub-adult II (two to three years old) Golden Eagles have mostly sub-adult feathers, with some retained juvenile feathers. Due to their extensive molt and wear at this age, **in sub-adult II birds the trailing edge of the wings typically appear more ragged than in other Golden Eagles**. On sub-adult II birds, **primary molt occurs toward the outer end of the wings**. On many Golden Eagles, the white patches on the wings are limited to the inner primaries, but birds with white patches limited to the secondaries only (or base of the wing) are sub-adult II or older. The central tail feathers of some sub-adult II birds may be replaced with adult feathers, causing the tail to look dark centered or split, a feature typical of sub-adult III birds.

Sub-adult III (three to four years old) Golden Eagles typically show a mix of adult and sub-adult tail feathers. The tail is dark with white patches on each side. When seen from below, this split-tailed appearance is only visible when the tail is spread. Sub-adult III Golden Eagles usually have dark remiges, but some show small white patches in the secondaries. Rarely, they retain a few juvenile secondaries (which may be longer) and look identical to sub-adult II birds underneath.

Sub-adult IV birds (four to five years old) often look identical to adults in the field. They typically lack white in the wings, but may show one or two small white spots along the secondaries. **The presence of white near the outer edge of the tail on sub-adult IV Golden Eagles is often the only way to tell them from adults.** From below, the white on the tail is only apparent when the tail is spread, and is not visible from above in many instances. **When seen gliding over-**

head in typical migratory fashion with the tail folded, it is impossible to precisely age many Golden Eagles, especially sub-adults and adults.

Adult Golden Eagles are completely dark underneath. The grayish banding on the flight feathers is all but impossible to see at a distance, but the wings of adults (and older sub-adults) may appear silvery underneath in direct sunlight similar to that of a Turkey Vulture. When illuminated by snow cover, the wings may appear almost whitish, like those of immature Golden Eagles. In direct sunlight, the sun's glare can cause the grayish tail bands on the topside of adults or the golden undertail coverts on all ages to appear whitish. However, they never look brilliant white as on immature birds. When adult Golden Eagles molt underwing coverts, they show small white patches in the wings similar to those of immature birds. However, these white patches appear irregular and are limited to the wing linings.

GE 01 - **Juvenile Golden Eagles** are dark underneath with white wing patches and a white base to the tail (top left). The white on the wings is extensive on some **juveniles** (top right) or lacking altogether on others (middle left). When backlit, the white on the tail is obvious on **juveniles** (middle right). In spring, **juveniles** may show signs of molt; note the shortened inner primary (bottom left). Golden Eagles at eye level or with folded tails may be difficult to age specifically (bottom right).

GE 02 - Sub-adult I (top left) and **sub-adult II** Golden Eagles (top right) often look identical to juveniles from below and are impossible to age specifically. However, birds with significant white on the wings and molting outer primaries are **sub-adult II birds** (middle left); note the ragged secondaries. **Sub-adult III** birds may have significant white in the wings that is limited to the secondaries (middle right), or show a tail with dark borders (bottom left). Birds with dark wings and white limited to the sides of the tail are typically **sub-adult III** (bottom right).

GE 03 - Sub-adult IV birds have minimal white on the sides of the tail (top left) or inner secondaries (top right). **Adults** are dark underneath, lacking white in the wings or tail (middle left). Over snow, **adult** remiges may appear somewhat whitish (middle right). Golden Eagles with dark wings and closed tails are impossible to age specifically; note the dark trailing edge and grayish tone to adult remiges and seemingly white tail base (bottom left). When shadowed or in a glide, Golden Eagles can be impossible to age (bottom right).

GE 04 - From above, **juveniles** show a white tail base and lack mottling on the upperwings (top left). A few **juveniles** show limited white on the tail (top right), while others show extensive white on the tail and primaries (middle left). In spring, **juveniles** show fading on the upperwing coverts (middle right) similar to the pale mottling of older birds (sub-adult I, bottom left). Sub-adult I and **sub-adult II** birds (bottom right) are often identical to each other on top and told from juveniles only by upperwing mottling. Note the outer primary on this sub-adult II not fully grown.

Shapes

Becoming familiar with structure (the shape) of a bird is a critical part of raptor field identification. The ability to differentiate the shapes of each raptor species, particularly wing shape, is often the key to identifying raptors in the field. Many species of raptor share similar plumages, especially at a distance or in the field. In these instances, structure is often the key feature that helps to differentiate among species. A combination of structural traits is more reliable than a single plumage trait in hawk identification.

Birders typically learn the silhouettes of soaring raptors first, since that is how they are frequently depicted in field guides and photographs. But birds on migration are often seen gliding at different altitudes and from a variety of angles. Because birds can change shape in an instant as they change posture, it is helpful to be familiar with each raptor in the various poses in which it is seen. One example is that most birds appear slimmer and longer-winged than usual when headed away or when approaching at eye level. Another aspect of structure is how the wing tips of each species trail farther beyond the secondaries in a steep glide as opposed to a shallow glide. Remember that each species normally holds its wings in a certain posture when soaring or gliding, but may hold them differently due to wind or landscape factors.

The following plates depict the various shapes and profiles that each species normally shows. Note the overall wing shape, body shape, head projection, tail length, and the subtle differences of each species in the various profiles, and the differences among these species. The plates are illustrated in black and white so that the focus is placed solely on structure, not plumage.

Sharp-shinned Hawk

Sharp-shinned Hawks have stocky, rounded wings and long, narrow tails. They are compact compared to other accipiters, with stockier wings, a slimmer tail, and a smaller head. Males are more compact than are females. In a glide or going away, the wing tips may appear pointed. Sharp-shinned Hawks hold their wings level when soaring and slightly drooped when gliding, but hunched-in compared to other accipiters.

Cooper's Hawk

Cooper's Hawks have stocky, rounded wings and long, narrow tails. They have relatively slightly slimmer wings and bodies and longer tails than do Sharp-shinned Hawks and Goshawks, but males can appear extremely similar in shape to Sharp-shinned Hawks. The head of Cooper's Hawks appears larger and sticks out farther than that of Sharp-shinned Hawks. Cooper's Hawks hold their wings in a slight dihedral or sometimes level when soaring and slightly drooped when gliding.

Northern Goshawk

Northern Goshawks have broad, rounded wings similar to those of Sharp-shinned Hawks, but they taper more toward the tips, making them quite angular along the back edge. The body is stout and the tail is long. Goshawks, especially adult males, appear falcon-like in a glide, but their wings are stockier overall and taper less at the tips than those of falcons. Goshawks hold their wings fairly flat when soaring and slightly drooped when gliding.

Northern Harrier

Northern Harriers have long, narrow wings and tails. The head is small, and the body is slim. Males have slightly shorter wings and tails than do females. In a glide, Harriers are similar in shape to Peregrine Falcons, but their wings are slimmer at the base and taper less at the "hands," they lack a tapered tail tip, and their heads and bodies are slimmer. Harriers hold their wings in a strong dihedral when soaring, a modified dihedral when gliding, and sometimes drooped on ridge updrafts.

Red-shouldered Hawk

Red-shouldered Hawks have somewhat long, broad wings that lack an obvious bulge and taper toward the tips but are more squared than those of other buteos. The body is stout and the tail is somewhat long and broad. In a glide, the wings are stocky and the wing tips protrude less than on other buteos except for Broad-winged Hawks. Red-shouldered Hawks soar with a slight dihedral or on flat wings and glide on drooped wings.

Broad-winged Hawk

Broad-winged Hawks have short, broad wings compared to those of other buteos, with pointed wing tips. The body is stout, the head is large, and the tail is somewhat short but very narrow when folded, appearing accipiter-like. In a glide, the wings are stocky and the wing tips barely protrude past the back of the wings except in a very steep glide. Broad-winged Hawks soar on flat wings and glide on bowed wings.

Swainson's Hawk

Swainson's Hawks have long, somewhat broad wings that are pointed in all postures. The body is stout and the tail is somewhat slim for a large buteo. In a glide, the wings are long and somewhat narrow; the wing tips protrude well past the back edge of the wings, making them appear "M"-shaped overhead. Swainson's Hawks soar with a strong dihedral and glide on bowed wings.

Red-tailed Hawk

Red-tailed Hawks have long, broad, bulging wings. The body is stout and the tail is broad. They soar with a moderate dihedral and glide on slightly bowed wings. In a glide, the wings are stocky and the wing tips protrude less beyond the back of the wings than those of other large buteos. Juveniles have slimmer wings and longer tails than do adults.

Ferruginous Hawk

Ferruginous Hawks have long, somewhat broad, angular, pointed wings that lack a noticeable bulge. The body is very stout and the tail is long and broad. In a glide, the wings are lengthy and the wing tips protrude well past the back edge of the wings, similar to the appearance of Swainson's Hawks. Ferruginous Hawks soar with a strong dihedral and glide with a modified dihedral.

Rough-legged Hawk

Rough-legged Hawks have long, somewhat broad wings, a fairly long tail, and a plump chest. In a glide, the wings are stocky but the wing tips protrude well past the back of the wings. Rough-legged Hawks soar with a moderate to strong dihedral and glide with a modified dihedral. Males are slimmer overall than are females, appearing somewhat Harrier-like in shape.

American Kestrel

American Kestrels have long, slim, pointed wings and long, narrow tails that do not taper toward the tip. The wings of Kestrels often appear smoothly curved and blunt at the tips compared to those of other falcons. Kestrels soar on flat wings and glide on slightly bowed wings. The wings are often hunched in when gliding into strong winds.

Merlin

Merlins have long, narrow, pointed wings that are relatively stockier compared to those of other falcons. Merlins have slightly shorter, broader tails, broader chests, and more sharply pointed wing tips compared to those of Kestrels. Merlins soar on flat wings and glide on slightly bowed wings

Peregrine Falcon

Peregrine Falcons have long, narrow, pointed wings and long, somewhat broad tails that often taper toward the tip when closed. The wings of Peregrines are slightly broader at the base than those of Kestrels and Prairie Falcons. The chest is broad and the head is larger compared to that of Kestrels and Merlins. Peregrines soar on flat wings and glide on slightly bowed wings.

Prairie Falcon

Prairie Falcons have long, narrow, pointed wings and long tails that sometimes taper toward the tip when closed. They are extremely similar in shape to Peregrines, but have slightly narrower wings, appearing Kestrel-like at times. The wings of Prairie Falcons sometimes appear blunt at the tips compared to those of Peregrines. The head, chest, and tail are broad compared to those of Kestrels and Merlins. Prairie Falcons soar on flat wings and glide on slightly bowed wings.

Black Vulture

Black Vultures have short, stocky wings with squared-off "hands," very short tails, and small heads. They hold their wings flat, with a slight dihedral, or slight modified dihedral in all postures. Black Vultures sometimes resemble Turkey Vultures or Bald Eagles, but are stockier overall with considerably shorter tails.

Turkey Vulture

Turkey Vultures have long, broad, straight-cut wings, and long, broad tails. The head is small, and the body is stout. They hold their wings with a strong dihedral or modified dihedral in all postures. Turkey Vultures can resemble eagles, but have smaller heads and less tapered, squared-off wings.

Osprey

Ospreys have long, narrow, distinctly gull-like wings, and relatively slim bodies. In a glide, they appear distinctly "M"-shaped. Ospreys always hold their wings slightly drooped. When soaring, Ospreys can resemble Bald Eagles, but are slimmer overall.

Bald Eagle

Bald Eagles have extremely long, somewhat broad wings and large, protruding heads. They are similar in shape to Golden Eagles, but the wings of Bald Eagles taper (or pinch in) less toward the body. The wings of juvenile and sub-adult I Bald Eagles are slightly broader than those of older birds. Bald Eagles typically hold their wings flat or slightly drooped in all postures, but may show a slight dihedral when soaring.

Golden Eagle

Golden Eagles have extremely long, somewhat broad wings that taper toward the body. They show similarities to buteos and vultures, but exhibit larger heads and longer wings; however, they show relatively smaller heads than do Bald Eagles. Golden Eagles hold their wings in a dihedral when soaring and in a modified dihedral or slightly drooped when gliding.

Photo Credits

All photos were taken by the author except those noted below.

Doug Backlund
GY 02 - top

Aaron Barna
RT 07 - top left

Tony Beck
GY 01 - top right

Vic Berardi
BW 04 - top left
RL 04 - middle left
MK 01 - top left, top right, middle left, bottom right
MK 02 - top left, top right, middle left, middle right

Adam Hutchins
NG 02 - middle right

Mike Lanzone
MK 01 - middle right, bottom left

Tony Leukering
Black Vulture portrait

Steve Mlodinow
ZT 01 - top left

Michael O'Brien
ST_WT 01 - Top left, top right, middle left

Luke Ormand
GY 01 - top left

Rob Palmer
GY 01 - bottom right

Michael Shupe
NG 01 - bottom right
Broad-winged Hawk portrait
BW 02 - top right, middle left, middle right
BW 04 - middle right
RL 05 - bottom right
RL 07 - top right

Brian Sullivan
CC 01 - all 6 images
RS 03 - top left, middle left
MK 02 - Bottom left, bottom right
ZT 01 - top right, middle right, bottom right
OT 01 - Swallow-tailed Kite (top), Hook-billed Kite (middle)

Gerrit Vyn
GY 01 - middle right